DARWIN
AMONG THE
MACHINES

DARWIN
AMONG THE
MACHINES

The Evolution of
Global Intelligence

George B. Dyson

§ HELIX BOOKS

 ADDISON-WESLEY PUBLISHING COMPANY, INC.

Reading, Massachusetts · Menlo Park, California · New York · Don Mills, Ontario
Harlow, England · Amsterdam · Bonn · Sydney · Singapore · Tokyo
Madrid · San Juan · Paris · Seoul · Milan · Mexico City · Taipei

Many of the designations used by manufacturers and sellers to distinguish their products are claimed as trademarks. Where those designations appear in this book and Addison-Wesley was aware of a trademark claim, the designations have been printed in initial capital letters.

The woodcut on page 228, from *The Famous History of Frier Bacon*, 1679, appears by permission of The Huntington Library, San Marino, California.

Library of Congress Cataloging-in-Publication Data

Dyson, George, 1953-
    Darwin among the machines : the evolution of global intelligence / George B. Dyson.
        p.   cm.
    Includes bibliographical references and index.
    ISBN 0-201-40649-7
    1. Artificial intelligence.   2. Artificial life.   3. Neural networks (Computer science)   I. Title.
Q335.D97   1997   97-858
006.3--dc21   CIP

Jacket design by Robert Dietz
Text design by Kenneth J. Wilson
Set in 10-point Palatino by Carlisle Communications

1 2 3 4 5 6 7 8 9-MA-0100999897
First printing, April 1997

Anything can happen once.

—PHILIP MORRISON

CONTENTS

PREFACE

EDGE OF THE WORLD

This is a book about the nature of machines. It is framed as history but makes no claim to have separated the fables from the facts. Both mythology and science have a voice in explaining how human beings and technology arrived at the juncture that governs our lives today.

I have attempted, in my own life and in this book, to reconcile a love of nature with an affection for machines. In the game of life and evolution there are three players at the table: human beings, nature, and machines. I am firmly on the side of nature. But nature, I suspect, is on the side of the machines.

In November of 1972, at the age of nineteen, I built a small tree house on the shore of Burrard Inlet in British Columbia, and settled in. In winter I consumed books and firewood; in summer I explored the British Columbian and Alaskan coasts. The tree house, ninety-five feet up in a Douglas fir, was paneled with cedar I found drifting in Georgia Strait, split into boards whose grain spanned as many as seven hundred years.

During those tree house winters I had lots of time to think. It got dark at four in the afternoon, rained for days on end, and, when the ocean fog rolled in, the earth, but not the sky, was obscured. At odd, unpredictable moments I found myself wondering whether trees could think. Not thinking the way we think, but thinking the way trees think; say, two or three hundred years to form the slow trace of an idea.

I spent the summers working on a variety of boats. When running at night I preferred to take the midnight-to-daybreak watch. By three or four in the morning, I was alone with the trace of unseen landforms on the radar screen and the last hour or two of night. I sometimes left the helm and paced the decks. The world receded in a phosphorescent wake, while birds appeared as red or green phantoms in the glow of the running lights, depending on whether they took wing on the port or starboard side. I also found myself slipping down into the engine room for more than the obligatory check.

When you live within a boat its engine leaves an imprint, deeper than mind, on neural circuits first trained to identify the acoustic signature of a human heart. As I had sometimes drifted off to sleep in the forest canopy, boats passing in the distance, and wondered whether trees might think, so I sat in the engine-room companionway in the small hours of the morning, with the dark, forested islands passing by, and wondered whether engines might have souls. This question threads its way through the chapters of this book.

We are brothers and sisters of our machines. Minds and tools have been sharpened against each other ever since a scavenger's stone fractured cleanly and the first cutting edge was held in a hunter's hand. The obsidian flake and the silicon chip are struck by the light of the same campfire that has passed from hand to hand since the human mind began.

This book is not about the future. Where we are at present is puzzling enough. I prefer to look into the past, exercising the historian's privilege of selecting predictions that turned out to be right. The past is where we find answers to our questions: Who are we, and why? The future is where we see questions to which the answers are up to us.

Do we remain one species, or diverge into many?

Do we remain of many minds, or merge into one?

ACKNOWLEDGMENTS

Princeton University's Firestone Library, the largest open-stack collection in the world, is one of the few libraries that require a university identification card to get in. The job of guarding the turnstile at the entrance to the library must dull one's attention over the years, and I discovered in 1967 that by melting into the crowd of students flooding into the library at 8:30 in the morning, it was usually possible to sneak in. Firestone's fifty-five miles of books, most of them shelved underground, offered a warm, anonymous refuge until it was safe to reappear out on the street and meet up with friends who had suffered through a day at school. I was left with a love of libraries, and a fear of librarians, that has lasted ever since.

Western Washington University's Fairhaven College granted me research associate status, with library privileges, to write this book. The Mabel Zoe Wilson Library is a small, comfortable facility, and to its resources I owe most of the citations appearing here. Special thanks go to Frank Haulgren and colleagues at interlibrary loan, who successfully pursued obscure requests. Bob Christenson, who enjoys confronting librarians as avidly as I shy away from them, helped excavate many things. Robert Keller, Marie Eaton, and others at Fairhaven College managed to bend the university's rules around my absence of credentials. Without such support this book would not exist.

The engines of evolution are driven by the recombination of genes; human creativity is driven by the recombination of ideas; literature is driven by the recombination of books. This book owes its elements to many others, cited elsewhere, and to two books that deserve special mention here. My father's *Origins of Life*[1] and my mother's *Gödel's Theorems*[2] contributed substantially to whatever limited understanding of the foundations of biology and of the foundations of mathematics is represented in this book. Both critiqued the manuscript as it took form, but any remaining errors or misinterpretations are my own.

In 1981 my sister, Esther Dyson, became editor of the *Rosen Electronics Letter*, a Wall Street investment newsletter that sensed wider implications as the personal-computer revolution began. Esther observed the new industry, and I observed Esther. All my perspectives on computational ecology can be traced to the *Rosen Electronics Letter* (which became *RELease 1.0* in 1983). This does not imply that Esther agrees with any of my interpretations of her work.

Thanks to Esther, I met literary agent John "No Wasted Motion" Brockman in 1984, who, nine years later, with Katinka Matson, helped precipitate this book. William Patrick at Addison-Wesley accepted an ambiguous proposal, and Jeff Robbins had the patience to await a manuscript, followed by the efficiency as editor to produce a book without additional delay. Others, including Danny Hillis, William S. Laughlin, James Noyes, Patrick Ong, and Ann Yow, offered encouragement at different stages along the way. The builders of my boat designs kept me afloat. I owe the last sentence in this book, and more, to David Brower—archdruid, mountaineer, and editor of landmarks from *In Wildness . . .* to *On the Loose*.

My daughter Lauren had just turned five, in 1994, when we watched a videotape describing Thomas Ray's digital organisms, self-reproducing numbers that had enraptured their creator by evolving new species and new patterns of behavior overnight. Ray was speaking at the Institute for Advanced Study, in Princeton, New Jersey, where forty years earlier the first experiments at evolving numerical organisms were performed. Ray's Tierran creatures inhabit a landscape entirely foreign to our own. Their expanding digital universe was first wrested into existence, out of the realm of pure mathematics, by the glow of twenty-six hundred vacuum tubes that flickered briefly at the dawn of digital programming in a low brick building at the foot of Olden Lane. Tom Ray and his portable universe now stood on ancestral ground.

"This is Tom Ray and his imaginary creatures," I said, explaining what we were watching partway through the tape. "But Dad," my daughter corrected, "they're *not* imaginary!"

She's right.

1

LEVIATHAN

Canst thou draw out leviathan with an hook?
or his tongue with a cord which *thou lettest down?*

Canst thou put an hook into his nose?
or bore his jaw through with a thorn?

Will he make many supplications unto thee?
will he speak soft words *unto thee?*

Will he make a covenant with thee?
wilt thou take him for a servant for ever?

Wilt thou play with him as with *a bird?*
or wilt thou bind him for thy maidens?

Shall the companions make a banquet of him?
shall they part him among the merchants?

Canst thou fill his skin with barbed irons?
or his head with fish spears?

Lay thine hand upon him, remember the battle, do no more.

—JOB 41:1–8

"Nature (the Art whereby God hath made and governes the World) is by the *Art* of man, as in many other things, so in this also imitated, that it can make an Artificial Animal," wrote Thomas Hobbes (1588–1679) on the first page of his *Leviathan; or, The Matter, Forme, and Power of a Common-wealth Ecclesiasticall and Civill,* published to great disturbance in 1651. "For seeing life is but a motion of Limbs, the beginning whereof is in the principall part within; why may we not say that all *Automata* (Engines that move themselves by

1

springs and wheeles as doth a watch) have an artificiall life?"[1] Hobbes believed that the human commonwealth, given substance by the power of its institutions and the ingenuity of its machines, would coalesce to form that Leviathan described in the Old Testament, when the Lord, speaking to Job out of the whirlwind, had warned, "Upon earth there is not his like, who is made without fear."

Three centuries after Hobbes, automata are multiplying with an agility that no vision formed in the seventeenth century could have foretold. Artificial intelligence flickers on the desktop and artificial life has become a respectable pursuit. But the artificial life and artificial intelligence that so animated Hobbes's outlook on the world was not the discrete, autonomous mechanical intelligence conceived by the architects of digital processing in the twentieth century. Hobbes's Leviathan was a diffuse, distributed, artificial organism more characteristic of the technologies and computational architectures approaching with the arrival of the twenty-first.

"What is the *Heart*, but a *Spring;* and the *Nerves*, but so many *Strings;* and the *Joynts*, but so many *Wheeles*, giving motion to the whole Body, such as was intended by the Artificer?" asked Hobbes. "*Art* goes yet further, imitating that rationall and most excellent worke of Nature, *Man*. For by Art is created that great LEVIATHAN called a COMMON-WEALTH . . . which is but an Artificiall Man."[2] Despite his reasoned arguments Hobbes was variously condemned by the monarchy, the Parliament, the universities, and the church. Hobbes saw human society as a self-organizing system, possessed of a life and intelligence of its own. Power was vested by mutual consensus, but not by divine right, in the hands of an assembly or a king. Loyalty was useful but need not be absolute. This ambivalence was viewed with suspicion from both sides. "Mr. Hobbs defyeth the whole host of learned men," and was "dangerous to both Government and Religion," warned Alexander Ross in *Leviathan Drawn out with a Hook,*[3] the first of a series of attacks that culminated with the citing by the House of Commons of Hobbes's blasphemies as a probable cause of the great fire and plague of 1666. Although threats against Hobbes were never executed, he destroyed his more incriminating manuscripts, fearing the worst. In his *Historical Narration Concerning Heresie, and the Punishment Thereof,* written in 1668, Hobbes maintained that his ideas did not fit the existing definition of heresy and accusations against him were unjust; in any event, he argued, there was no legal authority for burning heretics at the stake. Nonetheless, after Hobbes was safely dead, a decree by the University of Oxford in 1683 recommended that *Leviathan,* among other "Pernicious Books and Damnable Doctrines," be burned.[4]

Hobbes's blasphemy was his vision of a diffuse intelligence that was neither the supreme intelligence of God nor the individual intelligence of the human mind. Leviathan was a collective organism, transcending the individual beings and institutional organs of which it was composed. Human society, taken as a whole, constituted a new form of life, explained Hobbes, "in which, the *Soveraignty* is an *Artificiall Soul*, as giving life and motion to the whole body; The *Magistrates*, and other *Officiers* of Judicature and Execution, Artificiall *Joynts*; *Reward* and *Punishment* (by which fastned to the seate of the Soveraignty, every joynt and member is moved to performe his duty) are the *Nerves*, that do the same in the body Naturall; The *Wealth* and *Riches* of all the particular members, are the *Strength*; *Salus Populi* (the *peoples safety*) its *Businesse*; *Counsellors*, by whom all things needfull for it to know, are suggested unto it, are the *Memory*; *Equity* and *Lawes*, an artificiall *Reason* and *Will*; *Concord*, *Health*; *Sedition*, *Sicknesse*; and *Civill war*, *Death*."[5]

Hobbes sought not to diminish the intelligence of any existing being, human or divine, but rather to discover evidence of intelligence in the vacuum that supposedly intervened. As he argued against the physical vacuum demonstrated by the air pump of Robert Boyle, so he argued against the metaphysical vacuum that separated God from man. Hobbes hinted at a science of complex systems as comprehensive (and potentially heretical) as the two new sciences by which Galileo, befriended by Hobbes in 1636, had revealed the relative motion of all things. Hobbes's shortcomings as a mathematician, ridiculed by other natural philosophers, were outweighed by his facility with words. His ambition—when not distracted by civil war, the Restoration, or other social upheavals of the time—was to construct a consistent and purely materialistic natural philosophy of mind. "Motion produceth nothing but motion," he argued.[6] "And consequently every part of the Universe, is Body, and that which is not Body, is no part of the Universe: And because the Universe is All, that which is no part of it, is *Nothing*."[7] His analysis revealed deep-seated contradictions within the doctrines of the church. "Wee are told, there be in the world certaine Essences separated from Bodies, which they call *Abstract Essences, and Substantiall Formes*: For the Interpretation of which *Jargon*, there is need of somewhat more than ordinary attention. . . . Being once fallen into this Error of *Separated Essences*, they are thereby necessarily involved in many other absurdities that follow it. . . . Can any man think that God is served with such absurdities?"[8]

Hobbes protested strongly against the metaphysics of René Descartes (1596–1650). His objections, along with a terse response, were published in 1641 as an appendix to Descartes's *Meditationes de prima philosophia,* translated into English as *Six Metaphysical Meditations; Wherein it is Proved that there is a God. And that Mans Mind is really distinct from his Body.* "The question may be put *infinitely,* how do you *know* that you *know,* that you *know,* that you *know?* &c," argued Hobbes. "Wherefore . . . we cannot separate *thought* from *thinking matter,* it seems rather to follow, that a *thinking thing* is *material,* than that 'tis *immaterial.*"[9] Hobbes countered all the arguments that would reappear much later as arguments against the possibility of mind among machines. "*Ratiocination* will depend on *Words, Words* on *Imagination,* and perhaps *Imagination* as also *Sense* on the *Motion of Corporeal Parts;* and so the *Mind* shall be nothing but *Motions* in some Parts of an *Organical Body,*" he explained, treading dangerously close to heresy, though failing to dissuade Descartes.[10]

In suggesting, as Alexander Ross put it, "that our natural reason is the word of God" and that "it was a winde, not the holy spirit which in the Creation moved on the waters,"[11] Hobbes raised an upheaval that reverberated for three hundred years. The seeds of the Darwinian revolution, with all its ensuing controversies, were sown by Hobbes. The precedent for Bishop Samuel Wilberforce versus Thomas Huxley and Charles Darwin in 1860 was set in 1658 by Bishop John Bramhall versus Thomas Hobbes, launched with a sweeping salvo titled *The Catching of the Leviathan, or the Great Whale, Demonstrating out of Mr. Hobbs his own Works, That no man who is thoroughly an Hobbist, can be a good Christian, or a Good Common-wealths man, or reconcile himself to himself, Because his Principles are not only destructive to all Religion but to all Societies; extinguishing the Relation between Prince and Subject, Parent and Child, Master and Servant, Husband and Wife; and abound with palpable contradictions.*

Hobbes bore these attacks without flinching and made few concessions to the authorities of his time. He was famous for his irreverences, including an opinion that "the Episcopalians ridiculed the Puritans, and the Puritans the Episcopalians; but . . . the Wise ridiculed both alike."[12] Charles II, then the sixteen-year-old Prince of Wales, had been tutored by Hobbes while exiled in Paris in 1646; with the restoration of the monarchy in 1660 he invited Hobbes into his court. The king awarded Hobbes a small pension and gave him a measure of protection against his enemies, describing him as "a bear, against whom the Church played their young dogs, in order to exercise them."[13] The insults against Hobbes grew bolder on his death, such as the anonymous *Dialogues of the Living and the Dead,* which appeared in 1699, satirizing Hobbes as "a parcel of atoms jumbled together by chance." Hobbes had prepared for a protracted battle, leveling his own broadsides at his opponents, epitomized by his *Considerations upon the Reputation, Loyalty, Manners, & Religion, of Thomas Hobbes of Malmsbury, written by himself, by way of a Letter to a Learned Person*, in which he asked: "What kind of Attribute I pray you is *immaterial*, or *incorporeal* substance? Where do you find it in the Scripture? Whence came it hither, but from *Plato* and *Aristotle*, Heathens, who mistook those thin Inhabitants of the Brain they see in sleep, for so many *incorporeal* men; and yet allow them motion, which is proper only to things *corporeal?* Do you think it an honour to God to be one of these?"[14]

Hobbes advocated neither the pantheism of the ancients nor the atheism of which he was accused. He believed life and mind to be natural consequences of matter when suitably arranged; God to be a corporeal being, of perhaps infinitely higher mental order but composed of substance nonetheless; and damnation, to those so afflicted, a temporary state. The eloquence of his arguments wounded his critics deeply, whereas Hobbes suffered only superficially from the charges of heresy and promises of eternal hellfire pressed against him in response. "In writing books just as in real life," he wrote to Cosimo de' Medici in 1669, "enemies are more useful than friends."[15] *Leviathan* circulated widely, reprinted by underground or offshore press. "To my bookseller's, for 'Hobbs's Leviathan,'" noted Samuel Pepys in 1668, "which is now mightily called for; and what was heretofore sold for 8*s*. I now give 24*s*. at the second hand, and is sold for 30*s*., it being a book the Bishops will not let be printed again."[16]

Hobbes was a lifelong pacifist, a disposition he attributed to a premature birth precipitated by anxiety over the approach of the Spanish Armada in 1588. He meticulously cultivated new ideas and distilled them into words. "He walked much and contemplated," wrote his contemporary John Aubrey, "and he had in the head of his Staffe a pen and inke-horn, carried always a Note-book in his pocket, and as soon as a notion darted, he presently enterd it into his Booke, or els he should perhaps have lost it."[17] He played tennis until the age of seventy-five ("this he did believe would make him live two or three yeares the longer") and served up a lively game of words until silenced by a peaceful death at the age of ninety-one. "Neither the timorousness of his Nature from his Infancy, nor the decay of his Vital Heat in the extremity of old age," reported Aubrey, "chilled the briske Fervour and Vigour of his mind, which did wonderfully continue to him to his last."[18] His most outspoken critics were among the first to grant his intellect their respect. Steven Shapin and Simon Schaffer concluded their exhaustive study of the argument between Hobbes and Robert Boyle with an unambiguous judgment: "Hobbes was right."[19] Hobbes's vision was never so much extinguished as transformed.

Two centuries after Hobbes, the French electrodynamicist André-Marie Ampère sought to categorize all branches of human knowledge in his *Essay on the Philosophy of Science, or Analytic exposition of a natural classification of human knowledge.*[20] Reaching the field of political science through territory first explored by Hobbes (who composed *Leviathan* during his exile in Paris, before the French clerical authorities grew agitated by his ideas), Ampère coined a word with a far-reaching destiny: *Cybernétique.* Derived from Greek terminology referring to the steering of a ship, Ampère's *Cybernétique* encompassed that body of theory, complementary to but distinct from the theory of power, concerned with the underlying processes that direct the course of organizations of all kinds. In the second, posthumous volume of Ampère's *Essay,* published by his son in 1843, Ampère explains how he came to recognize a field of knowledge "which I name *Cybernétique,* from the word χυβερνετική, which was applied first, in a restricted sense, to the steering of a vessel, and later acquired, even among the Greeks, a meaning extending to the *art of steering in general.*"[21]

Ampère, an early advocate of the electromagnetic telegraph and mathematical pioneer of both game theory and electrodynamics, thereby anticipated the *Cybernetics* of Norbert Wiener, who, another century later, reinvented both Ampère's terminology and Hobbes's philosophy in their current, electronic form. "Although the term *cybernetics* does not date further back than the summer of 1947," wrote Wiener in 1948, "we shall find it convenient to use in referring to earlier epochs of the development of the field."[22] Wiener, who was involved in the development of radar-guided anti-aircraft fire control, which marked the beginning of rudimentary perception by electronic machines, was unaware until after the publication of *Cybernetics* of the coincidence in choosing a name coined by the same Ampère we now honor in measuring the flow of electrons through a circuit. In 1820, by demonstrating that electric currents are able to convey both power *and* information, Ampère had laid the foundations for Wiener's cybernetic principles of feedback, adaptation, and control.

We live in an age of embodied logic whose beginnings go back to Thomas Hobbes as surely as it remains our destiny to see new Leviathans unfold. Hobbes established that logic and digital computation share common foundations, suggesting a basis in common with mind. "Per ratiocinationem autem intelligo computationem," declared

Hobbes in 1655, or, "by *ratiocination*, I mean *computation*. Now to compute, is either to collect the sum of many things that are added together, or to know what remains when one thing is taken out of another. *Ratiocination*, therefore is the same with *Addition* or *Substraction;* and if any man adde *Multiplication* and *Division*, I will not be against it, seeing Multiplication is nothing but Addition of equals one to another, and Division nothing but a Substraction of equals one from another, as often as is possible. So that all Ratiocination is comprehended in these two operations of the minde, Addition and Substraction."[23]

This statement launched an argument far from settled after 340 years: If reasoning can be reduced to arithmetic, which, even in Hobbes's time, could be performed by mechanism, then is mechanism capable of reasoning? Can machines think? (Or, as Marvin Minsky put it, "Why do people think computers can't?")[24] Hobbes, the patriarch of artificial intelligence, was succeeded in this line of questioning by the young German lawyer and mathematician Gottfried Wilhelm von Leibniz (1646–1716), who made the first attempt at a system of symbolic logic and the first suggestion of a binary computing machine. The holy grail of capturing intelligence within a formal, mechanical system, however, slipped through Leibniz's grasp.

Or did it? The binary arithmetic and logical calculus of Leibniz and Hobbes's vague notions of reason as a mathematical function are now executed millions of times per second by thumbnail-size machines. Our formalization of logic is embedded microscopically in these devices, and by every available means of digital communication, from fiber optics to circulating floppy disks, the kingdom of the microprocessor is building a collective body of results. Philosophers and mathematicians have made limited progress at deconstructing the firmament of mind from the top down, while a grand, bottom-up experiment at building intelligence from elemental bits of addition and subtraction has been advancing by leaps and bounds. The results have more in common with the diffuse intelligence of Hobbes's Leviathan than with the localized artificial intelligence, or AI, that has now been promised for fifty years.

Is intelligence a formal (or mathematically definable) system? Is life a recursive (or mechanically calculable) function? What happens when you replicate discrete-state microprocessors by the billions and run these questions the other way? (Are formal systems intelligent? Are recursive functions alive?) Life and intelligence have learned to operate on any number of different scales: larger, smaller, slower, and faster than our own. Biology and technology evidence parallel tendencies toward collective, hierarchical processes based on information exchange. As information is distributed, it tends to be represented (encoded) by increasingly economical (meaningful) forms. This evolutionary process, whereby the most economical or meaningful representation wins, leads to a hierarchy of languages, encoding meaning on levels that transcend comprehension by the system's individual components—whether genes, insects, microprocessors, or human minds.

Binary arithmetic is a common language understood by switches of all kinds. The global population of integrated circuits—monolithic networks of microscopic switches that take only billionths of a second to switch between off and on—is growing by more than 100 million units per day.[25] Production of silicon wafer, approximately 2.5 billion square inches for the year 1994, is expected to double by the year 2000—enough raw material, to use an existing benchmark, for 30 billion Pentium microprocessors, of 3.3 million transistors each.[26] Intel's Pentium microprocessors are now manufactured, tested, and packaged at a cost of less than forty dollars each, while 350,000-transistor 486SXL embedded microprocessors cost less than eight dollars to manufacture and sell in quantity for about fifteen dollars each.[27] Microcontrollers—specialized microprocessors embedded in all kinds of things—were produced at a rate of more than 8 million units per day in 1996.[28] Over 200,000 non-embedded 32-bit microprocessors per day were shipped in 1995, and worldwide sales of personal computers exceeded 70 million units for the year. But the distinction between microprocessors and microcontrollers is increasingly obscure. Embedded devices are being integrated into the computational landscape, while computers are reaching beyond the desktop to become more deeply embedded in the control of all aspects of our world.

This digital metabolism is held together by telecommunications, spanning distance, and by memory, spanning time. Annual production of dynamic random-access memory (DRAM) now exceeds 25 billion megabits, and the manufacturing cost of 16-megabit memory circuits dropped below $10.00, or $0.62 per megabit, in 1996.[29] More than 100 million hard disk drives—averaging 500 megabytes each—were shipped in 1996. The market for electronic connectors now exceeds 20 billion dollars a year. Long-distance transmission of data has exceeded transmission of voice since 1995, with current telecommunications standards allowing the multiplexing of as many as 64,000 voice-equivalent channels over a single fiber optic pair.

Physicist Donald Keck, who wrote "Eureka!" in his Corning laboratory notebook after testing the first 200 meters of low-loss optical fiber in August 1970, estimated the worldwide installed base of optical fiber at more than 100 million kilometers at the end of 1996.[30] Eight million kilometers of telecommunications fiber were deployed in 1996 in the United States alone.[31] Much of this is "dark fiber" that awaits the growth of high-speed switching elsewhere in the global telecommunications network before it can be used. "The AT&T network is the world's largest computer," according to Alex Mandl of AT&T. "It is the largest distributed intelligence in the world—perhaps the universe," he claimed in 1995 (assuming that extraterrestrial civilizations have broken up their telecommunications industries into pieces smaller than AT&T).[32]

The emergence of life and intelligence from less-alive and less-intelligent components has happened at least once. Emergent behavior is that which cannot be predicted through analysis at any level simpler than that of the system as a whole. Explanations of emergence, like simplifications of complexity, are inherently illusory and can only be achieved by sleight of hand. This does not mean that emergence is not real. Emergent behavior, by definition, is what's left after everything else has been explained.

"Emergence offers a way to believe in physical causality while simultaneously maintaining the impossibility of a reductionist explanation of thought," wrote W. Daniel Hillis, a computer architect who believes that architecture and programming can only go so far, after which intelligence has to be allowed to evolve on its own. "For those who fear mechanistic explanations of the human mind, our ignorance of how local interactions produce emergent behavior offers a reassuring fog in which to hide the soul."[33] Although individual computers and individual computer programs are developing the elements of artificial intelligence, it is in the larger networks (or the network at large) that we are developing a more likely medium for the emergence of the Leviathan of artificial mind.

Sixty years ago, English logician Alan Turing constructed a theory of computable numbers by means of an imaginary discrete-state automaton, reading and writing distinguishable but otherwise intrinsically meaningless symbols on an unbounded length of tape. In Turing's universe there are only two objects in existence: Turing machine and tape. Turing's thought experiment was as close to Leibniz's dream of an elemental and universal language as mind, mechanism, or mathematics has been able to get so far. With the arrival of World War II, statistical analysis and the decoding of computable functions became a matter of life and death. Theory became hardware overnight. Turing and his wartime colleagues working for Allied intelligence at Bletchley Park found themselves coercing obstinate lengths of punched paper tape, at speeds of up to thirty miles per hour, through an optical mask linked by an array of photoelectric cells to the logical circuitry of a primitive computer named Colossus. Some fifteen hundred vacuum tubes, configured for parallel Boolean arithmetic, cycled through five thousand states per second, seeking to recognize a meaningful pattern in scrambled strings of code. The age of electronic digital computers was launched, secretively, as ten Colossi were brought on line by the time the war came to an end.

It has been nothing but Turing machines, in one form or another, ever since. Ours is the age of computable numbers, from the pocket calculator to Mozart on compact disc to the $89.95 operating system containing eleven million lines of code. We inhabit a computational labyrinth infested by billions of Turing machines, each shuffling through millions of internal states per second and set loose, without coordinated instructions, to read and write mutually intelligible strings of symbols on a communally unbounded, self-referential, and infinitely convoluted supply of tape.

Although our attention has been focused on the growth of computer networks as a medium for communication among human beings, beneath the surface lies a far more extensive growth in communication among machines. Everything that human beings are doing to make it easier to operate computer networks is at the same time, but for different reasons, making it easier for computer networks to operate human beings. Symbiosis operates by way of positive rewards. The benefits of telecommunication are so attractive that we are eager to share our world with these machines.

We are, after all, social creatures, formed by our nature into social units, as we ourselves are formed from societies of individual cells. Even H. G. Wells, who warned of a dark future as he approached the close of his life, held out hope for humanity through the globalization of human knowledge, described in his 1938 book *World Brain:* "In a universal organization and clarification of knowledge and ideas . . . in the evocation, that is, of what I have here called a World Brain . . . a World Brain which will replace our multitude of uncoordinated ganglia . . . in that and in that alone, it is maintained, is there any clear hope of a really Competent Receiver for world affairs. . . . We do not want dictators, we do not want oligarchic parties or class rule, we want a widespread world intelligence conscious of itself."[34] As we develop digital models of all things great and small, our models are faced with the puzzle of modeling themselves. As far as we know, this is how consciousness evolves.

Wells acknowledged memory not as an accessory to intelligence, but as the substance from which intelligence is formed. "The whole human memory can be, and probably in a short time will be, made accessible to every individual. . . . This new all-human cerebrum . . . need not be concentrated in any one single place. It need not be vulnerable as a human head or a human heart is vulnerable. It can be reproduced exactly and fully, in Peru, China, Iceland, Central Africa, or wherever else seems to afford an insurance against danger and interruption. It can have at once, the concentration of a craniate animal and the diffused vitality of an amoeba."[35] Writing from a perspective about midway, technologically, between the diffuse, largely unmechanized nature of Hobbes's Leviathan and the diffuse, highly mechanized information-processing structures of today, Wells held out the hope that this collective intelligence might improve on some of the collective stupidity exhibited by human beings so far. Let us hope that Wells was right.

Not everyone agrees that our great network of networks represents an emerging intelligence, or that it would be in our best interest if it did. Our intuitive association of intelligence with computational complexity has no precedent by which to grasp the combinatorial scale of the computer networks developing today. "Since the complexity is an exponential function of this kind of combinatorics, there is really a gigantic gap between computers and flatworms or any other simple kind of organism," warned Philip Morrison, considering the prospects for artificial intelligence in 1974. "Computer experts have a long, long way to go. If they work hard, their machines might approach the intelligence of a human. But the human species is not one person, it is 10^{10} of them, and that is entirely a different thing. When they tell you about 10^{10} computers, then you can start to worry."[36]

Those ten billion computers are not here yet, but the advance guard is settling in. Most are safely minding their own business, performing innocuous routines with no more intelligence than it takes to recalculate a spreadsheet, schedule a meeting, or adjust the ignition timing as you drive. Some are more visible than others, especially personal computers—microprocessors linked more or less intimately to the memories, intuitions, and decision-making abilities of individual human brains. Suddenly, with the convergence of the computer and telecommunications industries (not to mention the banking industry, which led the way) everything is being connected to everything else.

A circuit-switched communications network, in which real wires are switched to connect a flow of information between A and B, would be swamped by the intractable combinatorics of millions of computers demanding random access to their collective address space at once. All the switches in the world could never keep up. But with packet-switched data communications, collective computation scales gracefully as the number of processors (both electronic and biological) grows. Thanks to "hot-potato" routing algorithms, individual messages—the raw material from which intelligence is formed—are broken into smaller pieces, told where they are going but not how to get there, and reassembled after finding their own way to the destination address. Consensual protocols, running on all the processors in the net, maintain the appearance of robust connections between all the elements at once. The resulting free market for information and computational resources determines which connection pathways will be strengthened and which languish or die out. By the introduction of packet switching on an epidemic scale, the computational landscape is infiltrated by virtual circuitry, cultivating a haphazard, dendritic architecture reminiscent more of nature's design than of our own. Rules are simple, results complex. Does this signal the emergence of intelligence or merely the intellect of a bamboo forest growing toward the light?

Network architecture appears entirely random—as does, by coincidence or by design, the initial wiring of our own brains. Randomness has its reasons, however. "An argument in favor of building a machine with initial randomness is that, if it is large enough, it will contain every network that will ever be required," advised Irving J. Good, one of the pioneers of the Colossus, in a lecture on parallel processing given at IBM in 1958.[37] Whether growing a brain or evolving a telecommunications system, this seems to be good advice.

Computers may never embody mind at the level of human beings, despite a resurgence of such predictions every few years. But it is differences that make symbiotic relationships work. Symbiosis implies cooperation between distinguishable organisms, often a competition between host and parasite from which fruitful coexistence evolves. New and less distinguishable coalitions, such as lichens or eukaryotic cells, may be formed. "Life did not take over the globe by combat, but by networking," observed Lynn Margulis, describing how life evolved from the exchange of information between primitive chemical microprocessors the first time around.[38] Life began at least once and has been exploring its alternatives ever since. The cooperation between human beings and microprocessors is unprecedented, not in kind, but in suddenness and scale.

From simple congregations of simple molecules life moved, against all odds, to complex associations of complex molecules, forming a prolific molecular ecology eventually leading to living cells. Simple organisms were then established by associations of simple cells, followed by increasingly complex and differentiated cells forming increasingly complex and differentiated living forms. The social insects evolved elementary but highly successful collective organisms based on the behavior of individually simple parts, as Hobbes's *Leviathan* introduced the idea of an enduring collective organism composed of our own exceedingly complicated selves. And now, in the coalescence of electronics and biology, we are forming a complex collective organism composed of individual intelligences—governed not at the speed of Parliament but at the speed of light.

Is this the end of nature? Not by any means! Just as J. D. Bernal observed that "we are still too close to the birth of the universe to be certain about its death,"[39] so we are still too close to the beginning of nature (not to mention the beginning of science) to be certain about its end. As Hobbes's Leviathan sparked debate over the divine right of kings, so this new Leviathan signals an end to the illusion of technology as human beings exercising control over nature, rather than the other way around. The proliferation of microprocessors and the growth of distributed communications networks hold mysteries as deep as the origins of life, the source of our own intelligence, or the convergence of biology and technology toward a common language based on self-replicating strings of code. How can we imagine what comes next? As Loren Eiseley suggested concerning the possibility of life on other planets, in 1953, "It is as though nature had all possible, all unlikely worlds to make."[40]

Among the unlikely worlds that nature has yet to finish is the one that we call home. "And in this hope I return to my interrupted Speculation of Bodies Naturall," wrote Thomas Hobbes in the final paragraph of *Leviathan*, "wherein, (if God give me health to finish it,) I hope the Novelty will as much please, as in the Doctrine of this Artificiall Body it useth to offend."[41]

Nature, in her boundless affection for complexity, has begun to claim our creations as her own.

2

DARWIN AMONG THE MACHINES

As the vegetable kingdom was slowly developed from the mineral, and as in like manner the animal supervened upon the vegetable, so now in these last few ages an entirely new kingdom has sprung up, of which we as yet have only seen what will one day be considered the antedeluvian prototypes of the race. . . . as some of the lowest of the vertebrata attained a far greater size than has descended to their more highly organized living representatives, so a diminution in the size of machines has often attended their development and progress. . . . It appears to us that we are ourselves creating our own successors . . . giving them greater power and supplying by all sorts of ingenious contrivance that self-regulating, self-acting power which will be to them what intellect has been to the human race.

—SAMUEL BUTLER[1]

At the end of September 1859, a twenty-three-year-old Samuel Butler (1835–1902) sailed from England aboard the *Roman Emperor*, bound for Canterbury Settlement in New Zealand. The estranged son of Reverend Thomas Butler (rector of Langar, Nottinghamshire) and grandson of Dr. Samuel Butler (headmaster of Shrewsbury and bishop of Lichfield) was off to establish his independence as a sheep farmer in the New Zealand hills. Canterbury Settlement, founded by Church of England pilgrims granted title to "waste land in the middle island," was barely nine years old. Butler was renouncing the position (in church or college) his father expected of him, although he did not renounce some £4,400 in family capital that followed him from home. This and a degree in classics from Cambridge were the young emigrant's chief resources. He made good on both accounts.

"The world begins to feel very small when one finds one can get half round it in three months," wrote Butler during the voyage out.[2] On arrival at Port Lyttleton, near Christchurch, he purchased an experienced horse named Doctor, "a good river-horse, and very strong." Together they explored the surrounding territory, taking up a run of country at the divided headwaters of the Rangitata River (the homestead was christened Mesopotamia accordingly), where Butler put up a hut, established several thousand sheep, and lived happily until selling out his eight thousand acres at a substantial profit in 1864. Butler took well to the adventures along the way. The fellowship and solitude he found among the remote sheep stations is portrayed within the pages of his *First Year in Canterbury Settlement* in textures that suggest a Cambridge don making the rounds of his fellows' rooms. "After proceeding some few miles further I came to a station," he noted (13 February 1860) during his initial search for unclaimed territory, "where, though a perfect stranger, and at first (at some little distance) mistaken for a Maori, I was most kindly treated, and spent a very agreeable evening."[3] In March, after an excursion "in the extreme back country, and, positively, right up to a glacier," Butler and a fellow homesteader headed down from the hills: "We burnt the flats as we rode down, and made a smoke which was noticed between fifty and sixty miles off. I have seen no grander sight than the fire upon a country which has never before been burnt."[4]

New Zealand left its impression on Butler, and Butler left his impression on New Zealand. He adopted the landscape and anti-podean character of the remote colony as the model for his *Erewhon; or, Over the Range,* a satirical novel set in an isolated valley whose inhabitants had turned back the clock so as to preclude the development of intelligence among machines. Published anonymously in 1872, *Erewhon* was greeted with immediate success. "The reviewers did not know but what the book might have been written by a somebody whom it might not turn out well to have cut up, and whom it might turn out very well to have praised," Butler would later explain.[5] Unfortunately, "the demand fell off immediately on the announcement of my being the author,"[6] or, as Butler's friend and biographer Henry Festing Jones (1851–1928) put it, "as soon as *The Athenaeum* announced that *Erewhon* was by a nobody the demand fell 90 percent."[7]

Nonetheless, *Erewhon* made a name for Butler and provided the only measurable profit of his literary career, with 3,842 copies sold for a net gain of £69 3s. 10d. by his 1899 accounts.[8] Butler's father, who, according to H. F. Jones, "felt that success in this kind of literature was even more to be deprecated than success in any kind of painting,"[9] refused to read *Erewhon* and claimed that the book's appearance precipitated Mrs. Butler's death. Yet the world of the Erewhonians and the futility of their attempted sanctuary from machines remains as enduring a landmark as the New Zealand valley that bears the imprint Mesopotamia to this day.

Butler possessed an ability to find what he wanted in the world and to create life's accessories as he wished. He enjoyed art, so he took up painting, with enough success to exhibit at the Royal Academy. In honor of Handel, he took up composing and wrote, with Henry Festing Jones, a Handelian oratorio (*Narcissus*, 1888) as well as an album of gavottes, minuets, and fugues (Novello & Co., 1885). He cut a legendary figure in the New Zealand bush. "I shall never forget the small dark man with the penetrating eyes," remembered Sir Joshua Williams, "who took up a run at the back of beyond, carted a piano up there on a bullock dray, and passed his solitary evenings playing Bach's fugues; and who, when he emerged from his solitude and came down to Christchurch, was the most fascinating of companions."[10] Robert Booth, who hired on with Butler during Mesopotamia's second year, remembered him as "a literary man, and his snug sitting-room was fitted with books and easy chairs—a piano also. . . . Butler, Cook, and I would repair to the sitting-room, and round a glorious fire smoked or read or listened to Butler's piano. It was the most civilised experience I had had of up-country life."[11]

Erewhon and all the books that followed (until *Erewhon Revisited* in 1901) were published at Butler's own expense. The success of his nonfiction was uneven at best. A secure reputation had to await his autobiographical, anti-Victorian novel, *The Way of all Flesh*, which Butler left unpublished lest his relatives take offense. This courtesy was not extended to anyone else. "I have never written on any subject unless I believed that the authorities on it were hopelessly wrong," Butler proudly admitted.[12] His career was marked by a bitter dispute with Charles Darwin, precipitated by Darwin's failure to adequately credit the work of prior evolutionists, such as Georges Buffon (1707–1788), Jean-Baptiste Lamarck (1744–1829), Patrick Matthew (1790–1874), Robert Chambers (1802–1871), and Erasmus Darwin (1731–1802), who, in addition to being Charles's paternal grandfather, originated many of Charles's evolutionary ideas. Butler's criticism of Darwin's incomplete acknowledgments escalated into a sustained attack on the foundations of Charles Darwinism itself.

The ensuing controversy consumed four volumes of Butler's writings and twenty years of his life. On the advice of Thomas

Huxley, Darwin withheld comment, although, as Butler pointed out, some thirty-six references to "my" theory in the first edition of *Origin of Species* were deleted from subsequent editions of the book. Butler suffered the consequences of publicly attacking an intellectual hero of his time. "Has Mivart bitten him and given him Darwinophobia?" asked Huxley in a letter sent to Darwin in 1880. "It's a horrid disease and I would kill any son of a [Huxley leaves out a word but inserts a sketch] I found running loose with it without mercy."[13] The forces that Darwin and Huxley marshaled against their religious critics were directed against Butler's arguments as well—and with greater force, since Butler lacked institutional support.

In the absence of any understanding of how collectively nonrandom behavior can emerge, without the guidance of external (or natural) selection, from initially random events, Butler's interpretation of evolution appeared to be attempting a resurrection of the argument from design. In his rebellion against Darwinism (which Butler labeled *neo*-Darwinism, not to be confused with various neo-Darwinisms in circulation today) Butler was not retreating to the shelter of theology but thinking freely ahead for himself. "Butler's whole nature revolted against the idea that the universe was without intelligence," explained H. F. Jones.[14] The evolution of intelligence and the intelligence of evolution evidenced common principles of which life was at the same time both the cause and the result. Concerning the development of species, Butler wrote in *Luck, or Cunning?* that "they thus very gradually, but nonetheless effectually, design themselves."[15] Butler espoused a theory of species-level intelligence and grappled with the behavior of complex and self-organizing systems, as these lingering mysteries have more recently been framed. He favored Erasmus Darwin over Charles.

"From thus meditating on the great similarity of the structure of the warm-blooded animals, and at the same time of the great changes they undergo both before and after their nativity; and by considering in how minute a portion of time many of the changes of animals above described have been produced," asked Darwin, "would it be too bold to imagine, that in the great length of time, since the earth began to exist, perhaps millions of ages before the commencement of the history of mankind, would it be too bold to imagine, that all warm-blooded animals have arisen from one living filament, which THE GREAT FIRST CAUSE endued with animality, with the power of acquiring new parts, attended with new propensities, directed by irritations, sensations, volitions, and associations; and thus possessing the faculty of continuing to improve by its own inherent activity, and of delivering down those improvements by generation to its posterity, world without end!"[16] This was not Charles Darwin in his *Origin of Species* of 1859, but Erasmus Darwin in his *Zoonomia* of 1794. On one level, *Zoonomia; or, the Laws of Organic Life* was an encyclopedic medical text, a massive catalog of the great diversity of disease, contrasted by the appalling insufficiencies of what eighteenth-century medicine could do to help. On another level, *Zoonomia* was an attempt to construct "a theory founded upon nature, that should bind together the scattered facts of medical knowledge, and converge into one point of view the laws of organic life."[17]

"The great CREATOR of all things has infinitely diversified the works of his hands, but has at the same time stamped a certain similitude on the features of nature, that demonstrates to us, that *the whole is one family of one parent*," wrote Darwin in the preface to his book. "Shall we conjecture, that one and the same kind of living filaments is and has been the cause of all organic life?"[18] He elaborated further in the third edition of 1801: "I suppose that fibrils with formative appetencies, and molecules with formative aptitudes or propensities, produced by, or detached from, various essential parts of their respective systems, float in the vegetable or insect blood. . . . As these fibrils or molecules floated in the circulating blood of the parents, they were collected separately by appropriated glands of the male or female; and that finally on their mixture in the matrix the new embryon was generated, resembling in some parts the form of the father, and in other parts the form of the mother, according to the quantity or activity of the fibrils or molecules at the time of their conjunction."[19]

Erasmus (the father of fourteen children) was emphatic about the importance to genetic diversity of sex, noting that "if vegetables could only have been produced by buds and bulbs, and not by sexual generation, that there would not at this time have existed one thousandth part of their present number of species."[20] He developed some peculiar ideas about the effects of imagination on hereditary characteristics, offering clinical advice as to how "the sex of the embryon . . . may be made a male or a female by affecting the imagination of the father at the time of impregnation . . . but the manner of accomplishing this cannot be unfolded with sufficient delicacy for the public eye."[21] He focused his literary attentions, more or less discreetly, on the sexual life of plants and noted that the sexual exuberance of flowers reaches into the warm-blooded kingdom of, for instance, birds: "The final cause of this contest among the males seems to be, that the strongest and most active animal should propagate the species, which should thence become improved."[22]

Erasmus Darwin identified the essential principles of natural selection, descent with modification, and other pillars of evolutionary thought. "The great globe itself, and all that it inhabit, appear to be in a perpetual state of mutation and improvement," he noted in *The Temple of Nature; or, the Origin of Society* in 1803.[23] His evolutionary timescale was more realistic than that of his grandson Charles, and he was careful to emphasize that the study of evolution, rather than diminishing the power of God, served to glorify his work. "The world itself might have been generated, rather than created; that is, it might have been gradually produced from very small beginnings, increasing by the activity of its inherent principles, rather than by a sudden evolution of the whole by the Almighty fiat," he wrote in 1794. "What a magnificent idea of the infinite power of THE GREAT ARCHITECT! THE CAUSE OF CAUSES! PARENT OF PARENTS! ENS ENTIUM! For if we may compare infinities, it would seem to require a greater infinity of power to cause the causes of effects, than to cause the effects themselves."[24]

Erasmus Darwinism, however widely acclaimed at the time, has been obscured by a lingering confusion, perpetuated by both Charles Darwin and Samuel Butler, that equates the work of Erasmus Darwin with the errors of his follower Lamarck. A respected French naturalist and protégé of Buffon, Lamarck made lasting contributions to science, dividing the animal kingdom into vertebrates and invertebrates and assigning the label *biology* to the study of life. He is most famous, however, for his mistaken belief in the inheritance of acquired characteristics, the classic example being that giraffes grew taller by stretching their necks. Lamarckism reflected the prevailing views of the time and, indeed, was supported by Charles Darwin's provisional hypothesis of pangenesis, published in 1868. The views of Erasmus were in some respects less Lamarckian, and closer to the modern synthesis, than those expressed much later by Charles. But Erasmus failed to develop a concise packaging for his argument. He either published his observations as lengthy footnotes to his unwieldy poems or concealed them within his *Zoonomia*, expanded to fourteen hundred pages in the third edition of 1801. Sixty years later, Charles Darwin would be justly proclaimed a prophet, but, as Butler argued in *Evolution, Old and New*, he had inherited—not invented—the evolutionary faith.

From an otherwise modest position as surgeon of Lichfield, fifteen miles north of Birmingham, Dr. Erasmus Darwin became one of the foremost physicians of his time. Refusing the king's invitation to move to London, he kept up his daily rounds, dispensing his skills with uncommon generosity and moving as freely among all social circles as the bad state of the roads allowed. He lobbied prominently against the institution of slavery and for the humane treatment of the insane. Abstaining from both alcoholic spirits *and* Christianity, he embraced science and invention with an intellectual appetite exceeding his visible but less publicly celebrated appetites for female company and food. "Eat or be eaten," he is said to have advised his patients, following his own prescription to the extent that his dining table was modified to accommodate his girth. "In his youth Dr. Darwin was fond of sacrificing to both Bacchus and Venus," reported an anonymous contemporary, "but he soon discovered that he could not continue his devotions to both these deities without destroying his health and constitution. He therefore resolved to relinquish Bacchus, but his affection for Venus was retained to the last period of life."[25]

Erasmus Darwin was a ringleader of the Industrial Revolution, helping to spark the evolution of machines as surely as some unknown Cambrian ancestor of ours ignited the diversification of metazoan life. As Charles's son Francis Darwin (1848–1925) remarked, "Erasmus had a strong love of all kinds of mechanism, for which Charles Darwin had no taste."[26] In the 1760s, inspired by the Birmingham visits of Benjamin Franklin and drawing on his friendships with Matthew Boulton, Josiah Wedgwood, James Keir, William Small, and James Watt, Darwin founded the Lunar Society of Birmingham, an informal association of natural philosophers and industrialists whose meetings were scheduled to allow the full moon to assist its members home. The group of self-styled "Lunaticks" formed a nucleus for the industrialization of Britain, and either directly or via the interlocking relationships of the Lunar Society Erasmus Darwin had a hand in the origin of almost every species of mechanism explicit or implicit in the technologies of today.

Amid the peculiar triumphs and routine horrors of an eighteenth-century medical practice, Erasmus Darwin's notebooks contain rough sketches for pumps, steam turbines, horizontal-axis windmills, canal lifts, speaking machines, internal combustion engines, a compressed-air–powered ornithopter, a hydrogen–oxygen rocket motor, and even an automatic water closet that flushes itself when one opens the door to leave. Driven to inspiration during his tedious rounds ("I, imprison'd in a post-chaise, am joggl'd, and jostl'd and bump'd, and bruised along the King's highroad"),[27] Darwin proposed several improvements to horse-drawn carriages, although a misadventure with one of his prototypes in 1768 left him lame for the remainder of his life. Anticipating Samuel Butler, he owned a horse named Doctor, and with steam power on the horizon, he was for a time obsessed with the vision of a steam-driven "fiery chariot" that would replace the horse. "As I was riding Home yesterday," he wrote to Matthew Boulton, "I consid'd the Scheme of ye fiery Chariot—and ye longer I contemplated this favourite Idea, ye [more] practicable it appear'd to me."

"I am quite mad of this Scheme," Darwin continued, providing Boulton with a prospectus for a three-wheeled vehicle propelled by twin cylinders and an ingeniously differential rear-wheel drive. "By ye management of the steam cocks ye motion may be accelerated, retarded, destroy'd, revised, instantly & easyly. And if this answers in Practise as it does in theory, ye Machine can not fail of success." Boulton, the original pioneer of mass production (from belt buckles to steam engines), was too far in debt to act on Darwin's suggestion at the time, but the concept would resurface, like Darwinism, first in the age of railroads and then in the age of automobiles. A few years later, when James Watt developed the condenser engine, it was Darwin who promoted the Boulton & Watt partnership that brought the Industrial Revolution—and, soon enough, the "fiery chariot"—to life. Below Darwin's signature was appended a prophetic postscript: "I think four wheels would be better—adieu."[28]

Science fiction, as well as the automobile, owes Erasmus Darwin a founding credit. In a preface to the first (and anonymous) edition of Mary Wollstonecraft Shelley's *Frankenstein; or, the Modern Prometheus* (1818), Percy Shelley acknowledged that "the event on which this fiction is founded has been supposed by Dr. Darwin, and some of the physiological writers of Germany, as not of impossible occurrence."[29] In her introduction to the 1831 edition, Mary Shelley, who wrote the novel at age nineteen, also acknowledged Darwin, noting, "I speak not of what the Doctor really did, or said that he did, but, as more to my purpose, of what was then spoken of as having been done by him. . . . Perhaps a corpse would be re-animated; galvanism had given token of such things: perhaps the component parts of a creature might be manufactured, brought together, and endued with vital warmth."[30]

Darwin's electrotherapy treatments, widely noted in Shelley's time, still bring Dr. Frankenstein's experiments to mind. "Two thick brass wires, about 2 ft long, communicate from each extremity of the [Galvanic] pillar to each temple. The temples must be moistened with brine," wrote Darwin to the duchess of Devonshire in 1800. "The shock is so great as to make a flash in the eyes, and to be felt th[r]ough both the temples. . . . I have one patient here, a lady from near Scarborough, who has used it daily for giddyness with good success."[31] Darwin found that electric shocks could cure hepatic paralysis and renew the mobility of injured limbs. Luigi Galvani had shown the power of electric fluid to animate the legs of frogs; what additional powers might Darwin's experiments unleash? A notice in the Birmingham *Gazette* on 23 October 1762 invited anyone "whom the Love of Science may induce" to visit Dr. Darwin's laboratory: "The body of the Malefactor, who is order'd to be executed at Lichfield on Monday the 25th instant, will be afterwards conveyed to the house of Dr. Darwin, who will begin a Course of Anatomical Lectures, at Four o'clock on Tuesday evening, and continue them every Day as long as the Body can be preserved."[32]

"Dr. Darwin possesses perhaps a greater range of knowledge than any man in Europe," remarked Samuel Coleridge, who coined the word *Darwinising* in reference to evolutionary speculations; in this as in most other Darwinisms, Erasmus preceded Charles.[33] "The Darwinian theory of evolution is very much a family affair," concluded Desmond King-Hele, "in which the shares of Erasmus and his grandson Charles are more nearly connected, and more nearly equal, than is usually supposed."[34] Whether Charles's neglect of his grandfather's work was a conscious or unconscious oversight has been diagnosed both ways. The first edition of *Origin of Species* makes no mention of Erasmus Darwin. "The history of error is quite unimportant," explained Darwin to Huxley.[35] In the third edition, of 1861, Darwin added a "brief, but imperfect" historical sketch, in which he commented in a footnote that "it is curious how largely my grandfather, Dr. Erasmus Darwin, anticipated the erroneous grounds of opinion, and the views of Lamarck." This cast his grandfather in all but invisible type.

In 1879, Charles Darwin published, with a lengthy introduction, an English translation of Ernst Krause's *Life of Erasmus Darwin,* just as Butler was about to publish his *Evolution, Old and New.*[36] Instead of pacifying Butler, Darwin's belated acknowledgment of his grandfather had the opposite effect. Butler discovered that Darwin's translation of the original article by Krause, accompanied by "a guarantee for its accuracy" and presented as predating the appearance of *Evolution, Old and New,* contained several additional passages, including a final paragraph that Butler interpreted as a personal attack. "Erasmus Darwin's system was in itself a most significant first step in the path of knowledge which his grandson has opened up for us," suspiciously appended Krause, "but to wish to revive it at the present day, as has actually been seriously attempted, shows a weakness of thought and a mental anachronism which no one can envy."[37]

The Darwin–Butler dispute arose from an alliance gone awry. The grandson of the surgeon of Lichfield and the grandson of the bishop of Lichfield had been launched on a collision path, burdened by illustrious ancestors and driven to claim new territory for themselves. In the cold climate of a Victorian childhood the Reverend Thomas Butler is remembered as particularly harsh. Butler's alienation from his father and the church was followed by a disillusionment with Darwinism, which he denounced as early as January 1863 as "nothing new, but a *rechaufée.*"[38] Charles Darwin had been a student of Butler's grandfather and an acquaintance of Butler's father, who noted that "he inoculated me with a taste for Botany which has stuck by me all my life."[39] Darwin would only reciprocate with a comment that "nothing could have been worse for the development of my mind than Dr. Butler's school."[40]

Darwin's great treatise appeared in November 1859, but, recalled Butler, "being on my way to New Zealand when the *Origin of Species* appeared, I did not get it until 1860 or 1861."[41] The long sea voyage, the grand spectacle of the New Zealand wilderness, and a religious upbringing that sought to shift its convictions to a scientific faith rendered Butler keenly receptive to the theories presented in Darwin's book. Reading *Origin of Species* by candlelight in a thatched-roof hut, the constellations of the Southern Hemisphere above, Butler's imagination took flight beyond where Darwin left off. "Residing eighteen miles from the nearest human habitation, and three days' journey on horseback from a bookseller's shop, I became one of Mr. Darwin's many enthusiastic admirers," Butler recollected, "and wrote a philosophical dialogue (the most offensive form, except poetry and books of travel into supposed unknown countries, that even literature can assume) upon the *Origin of Species.*"[42]

This dialogue was printed anonymously in the Canterbury *Press* of 20 December 1862. By some means a copy reached Charles Darwin who, in forwarding it to an unknown editor in England, noted that "this Dialogue, written by some [one] quite unknown to Mr. Darwin, is remarkable from its spirit and from giving so clear and accurate a view of Mr. D[arwin]'s theory. It is also remarkable from being published in a colony exactly 12 years old, in which it might have [been] thought only material interests would have been regarded."[43]

Butler's dialogue aroused much discussion in the colony, and it was followed on 13 June 1863 by another installment, signed "Cellarius" and titled *Darwin Among the Machines.* In this essay Butler laid out the ideas that would be incorporated into *Erewhon* as the "Book of the Machines." "We find ourselves almost awestruck at the vast development of the mechanical world, at the gigantic strides with which it has advanced in comparison with the slow progress of the animal and vegetable kingdom," warned Butler. "We shall find it impossible to refrain from asking ourselves what the end of this mighty movement is to be. . . . The machines are gaining ground upon us; day by day we are becoming more subservient to them; more men are daily bound down as slaves to tend them; more men are daily devoting the energies of their whole lives to the development of mechanical life."[44]

Butler's essay did more than spoof a fashionable theory; it coupled a meticulous analysis of Darwin's thesis to a keenly unencumbered view of the world as it stood in 1863. On his return to London, Butler produced another commentary, "The Mechanical Creation," published in the (London) *Reasoner*, 1 July 1865. "Those who accept the Darwinian theory will not feel inclined to deny that whatever impulse the animal and vegetable kingdoms have sprung from, has been derived from within the natural influences which operate upon this world, and not from any extra natural source," argued Butler. "They will believe that the changes and chances with which countless millions of years have been pregnant, have brought the existing organizations to their present condition without any specially creative effort of an overruling mind. What shall we think then? That the resources of nature are at an end, and that the animal phase is to be the last which life on this globe is to assume? or shall we conceive that we are living in the first faint dawning of a new one? Of a life which in another ten or twenty million years shall be to us as we to the vegetable? What has been may be again, and although we grant that hardly any mistake would be more puerile than to individualize and animalize the at present existing machines—or to endow them with human sympathies, yet we can see no a priori objection to the gradual development of a mechanical life, though that life shall be so different from ours that it is only by a severe discipline that we can think of it as life at all."[45]

The relations between mind and mechanism have been argued since the time of Aristotle and Lucretius, the distinctions given a trademark presentation by René Descartes in his 1637 *Discourse touching the method of using one's reason rightly and of seeking scientific truth.* Butler adopted an open-minded position that "the theory that living beings are conscious machines, can be fought as much and just as little as the theory that machines are unconscious living beings; everything that goes to prove either of these propositions goes just as well to prove the other also."[46] This was less radical a view than that suggested by Darwin's colleague Thomas Huxley, who announced in 1870 that "we shall sooner or later arrive at a mechanical equivalent of consciousness, just as we have arrived at a mechanical equivalent of heat."[47]

In *Erewhon's* "Book of the Machines" the author of the anonymous manifesto presented within the anonymous book gives voice to these concerns: "Why may not there arise some new phase of mind which shall be as different from all present known phases as the mind of animals is from that of vegetables? It would be absurd to attempt to define such a mental state (or whatever it may be called), inasmuch as it must be something so foreign to man that his experience can give him no help towards conceiving its nature; but surely when we reflect upon the manifold phases of life and consciousness which have been evolved already, it would be rash to say that no others can be developed, and that animal life is the end of all things. There was a time when fire was the end of all things; another when rocks and water were so.... There is no security ... against the ultimate development of mechanical consciousness, in the fact of machines possessing little consciousness now.... Either, a great deal of action that has been called purely mechanical and unconscious must be admitted to contain more elements of consciousness than has been allowed hitherto (and in this case germs of consciousness will be found in many actions of the higher machines)—or (assuming the theory of evolution but at the same time denying the consciousness of vegetable and crystalline action) the race of man has descended from things which had no consciousness at all. In this case there is no *a priori* improbability in the descent of conscious (and more than conscious) machines from those which now exist."[48]

In May 1872, Butler sent a letter to Darwin apologizing "about a portion of the little book *Erewhon* which I have lately published, and which I am afraid has been a good deal misunderstood. I refer to the chapter upon Machines.... I am sincerely sorry that some of the critics should have thought that I was laughing at your theory, a thing which I never meant to do, and should be shocked at having done."[49] In reply, Darwin invited Butler to visit him at the Darwin estate at Down. Butler stayed with the Darwins for a weekend, a visit, wrote the Darwins' houseguest, "of which I shall always retain a most agreeable recollection."[50] It was the memory of this visit, perhaps, that would prompt Darwin to write to Huxley eight years later that "the [Butler] affair has annoyed and pained me to a silly extent ... until quite recently he expressed great friendship for me, and said he had learnt all he knew about evolution from my books."[51]

The Butler–Darwin quarrel smoldered for many years. Reconciliation was achieved only after both parties were deceased, the peace mediated between Francis Darwin on behalf of his father and Henry Festing Jones on behalf of Butler's literary estate.[52] The affair drew considerable attention at the time; what the nineteenth century lacked in television it made up for with a facility with words. "When a writer who has not given as many weeks to the subject as Mr. Darwin has given years," complained the *Saturday Review,* "is not content to air his own crude though clever fallacies, but assumes to criticize Mr. Darwin with the superciliousness of a young schoolmaster looking over a boy's theme, it is difficult not to take him more seriously than he deserves."[53]

"When I thought of Buffon, of Dr. Erasmus Darwin, of Lamarck, and even of the author [Robert Chambers] of the *Vestiges of Creation,* to all of whom Mr. Darwin had dealt the same measure which he was now dealing to myself," responded Butler, "when I thought of these great men, now dumb, who had borne the burden and heat of the day, and whose laurels had been filched from them ... dead men, on whose behalf I now fight, as I trust that some one—whom I thank by anticipation—may one day fight on mine."[54]

Was Samuel Butler right? Although the priority of Erasmus Darwin is now acknowledged, Butler's own evolutionary theories remain discredited as the unscientific speculations of a bitter and self-published crank. But several of the arguments he made in *Life and Habit* (1878), *Evolution, Old and New* (1879), *Unconscious Memory* (1880), and *Luck, or Cunning?* (1887) anticipated questions that are gnawing at the pillars of Darwinism today.

Butler's obsession with "the substantial identity between heredity and memory, and the reintroduction of design into organic development" anticipated the discovery of the genetic code and presaged the mysteries that bedevil our understanding of how living organisms are translated to and from sequential strings of DNA. The engines of evolution are driven by computational processes whose alphabet has been deciphered but whose language we do not yet understand. Butler's notion of species as composite organisms, transcending the temporal and physical boundaries between individuals, is echoed by recent models of how the space of evolutionary possibilities is searched and brought to life. The ghost of Samuel Butler haunts the fringes of evolutionary biology today. How random is random variation? Is life the work of natural selection alone—or is there an element of intelligent search, if not design?

In 1876, Butler explained the continuity of the germ plasm and hinted at what Richard Dawkins (1976) would label the *Selfish Gene:* "See the ova only and consider the second ovum as the first two ovas' means not of reproducing themselves but of continuing themselves—repeating themselves—the intermediate lives being nothing but, as it were, a long potato shoot from one eye to the place where it will grow its next tuber."[55] This insight would be immortalized as the aphorism that a chicken is an egg's way of making another egg. Butler's ideas about ideas, expressed best in his introduction to *Luck, or Cunning?* anticipated what, also thanks to Dawkins, we now call memes: "Ideas are like plants and animals in this respect also. I do not merely mean their growth in the minds of those who first advanced them, but that larger development which consists in their subsequent good or evil fortunes—in their reception, favourable or otherwise, by those to whom they were presented. This is to an idea what its surroundings are to an organism, and throws much the same light upon it that knowledge of the conditions under which an organism lives throws upon the organism itself."[56]

In examining the prospects for artificial intelligence and artificial life Butler faced the same mysteries that permeate these two subjects today. "I first asked myself whether life might not, after all, resolve itself into the complexity of arrangement of an inconceivably intricate mechanism," he recalled in 1880, retracing the development of his ideas. "If, then, men were not really alive after all, but were only machines of so complicated a make that it was less trouble to us to cut the difficulty and say that that kind of mechanism was 'being alive,' why should not machines ultimately become as complicated as we are, or at any rate complicated enough to be called living, and to be indeed as living as it was in the nature of anything at all to be? If it was only a case of their becoming more complicated, we were certainly doing our best to make them so."[57]

These questions can be distilled into one essential puzzle—the origin of life. "We wanted to know whence came that germ or those germs of life which, if Mr. Darwin was right, were once the world's only inhabitants," asked Butler. "They could hardly have come hither from some other world; they could not in their wet, cold, slimy state have travelled through the dry ethereal medium which we call space, and yet remained alive. If they travelled slowly, they would die, if fast, they would catch fire."[58] The only viable answer, without recourse to some higher being "at variance with the whole spirit of evolution," was that life "had grown up, in fact, out of the material substances and forces of the world"—as life might once again be growing up out of the material substances and forces of machines.

As Charles Darwin borrowed from his grandfather, I am now borrowing ideas that developed on family ground. My own father, Freeman J. Dyson, a mathematical physicist better known as one of the architects of quantum electrodynamics, took a midcareer detour into theoretical biology that resulted in a thin volume titled *Origins of Life*. The essence of my father's hypothesis was that life began not once, but twice. "It is often taken for granted that the origin of life is the same thing as the origin of replication," he wrote, noting that "it is important here to make a sharp distinction between replication and reproduction. . . . Cells can reproduce but only molecules can replicate. In modern times, reproduction of cells is always accompanied by replication of molecules, but this need not always have been so. . . . Either life began only once, with the functions of replication and metabolism already present in rudimentary form and linked together from the beginning, or life began twice, with two separate kinds of creatures, one kind capable of metabolism without exact replication, the other kind capable of replication without metabolism. . . . The most striking fact which we have learned about life as it now exists is the ubiquity of dual structure, the division of every organism into hardware and software components, into protein and nucleic acid. I consider dual structure to be prima facie evidence of dual origin. If we admit that the spontaneous emergence of protein structure and of nucleic acid structure out of molecular chaos are both unlikely, it is easier to imagine two unlikely events occurring separately over a long period of time."[59]

Over a period of twenty years, Dyson developed a toy mathematical model that "allows populations of several thousand molecular units to make the transition from disorder to order with reasonable probability."[60] These self-sustaining—and haphazardly reproducing—autocatalytic systems then provide energy (and information) gradients hospitable to the development of replication, perhaps first of parasites infecting the metabolism of primitive precursors of modern cells. Once metabolism is infected by replication, as the Darwins showed us, natural selection will do the rest.

Natural selection does not *require* replication; statistically approximate reproduction, for simple creatures, is good enough. The difference between replication (producing an exact copy) and reproduction (producing a similar copy) is the basis of a broad generalization: genes *replicate*, but organisms *reproduce*. As organisms became more complicated, they discovered how to *replicate* instructions (genes) that could help them reproduce; looking at it the other way around, as instructions became more complicated, they discovered how to *reproduce* organisms to help *replicate* the genes.

If organisms truly replicated, or reproduced even an approximate likeness of themselves without following a distinct set of inherited instructions, we would have Lamarckian evolution, with acquired characteristics transmitted to the offspring. According to the dual-origin hypothesis, natural selection may have operated in a purely statistical fashion for millions if not hundreds of millions of years before self-replicating instructions took control. This brings us back to Butler versus Darwin, because during this extended evolutionary prelude Lamarckian, not neo-Darwinian, selection would have been at work. We should think twice before dismissing Lamarck because Lamarckian evolution may have taken our cells the first—and most significant—step toward where we stand today. Genotype and phenotype may have started out synonymous and only later become estranged by the central dogma of molecular biology that allows communication from genotype to phenotype but not the other way. Life, however, arrives at distinctions by increments and rarely erases its steps. Remnants of Lamarckian evolution may be more prevalent, biologically, than we think—not to mention Lamarckian tendencies among machines.

"The experts were uniformly unenthusiastic," Freeman Dyson commented, describing how his venture into biology was received. "Roughly speaking, the difference of view between me and the community of experts is that the experts believe that RNA came first in the evolution of life whereas I believe that proteins came first.... The 'RNA world' has become an accepted dogma doubted only by a few heretics like me."[61]

My father asked three fundamental questions: "Is life one thing or two things? Is there a logical connection between metabolism and replication? Can we imagine metabolic life without replication, or replicative life without metabolism?"[62] These same three questions surround the origin(s) of life among machines. Here, too, a dual-origin hypothesis can shift the balance of probabilities in life's favor once the distinction between reproduction and replication is understood. In looking for signs of artificial life, either on the loose or cooked up in the laboratory, however permeable this distinction may prove to be, one should expect to see signs of metabolism without replication and replication without metabolism first. If we look at the world around us, we see a prolific growth of electronic metabolism, populated by virulently replicating code—just as the dual-origin hypothesis predicts.

The same dogma that has pervaded most theories of the origin of life—that life and replication are synonymous and arose simultaneously, however unlikely the event—has clouded the subject of artificial life. The first problem is to define what life, real or artificial, is. A prevailing assumption is that life begins with the genesis of self-replicating organisms, programs, or machines. Self-replication is a sufficient but by no means necessary condition for the origins or propagation of life. Replicators, when they make an appearance, will rapidly gain the upper hand, but this does not mean that they come first. Nor does it mean that replicators will thereafter keep the field to themselves. Under the neo-Darwinian regime—not so much a consequence of the origins of life as a consequence of the origins of death—replicators will, in the long run, win. But there is no law against changing the rules. Intelligence and technology are bringing Lamarckian mechanisms into play, with results that may leave the slow pace of Darwinian trial and error behind.

"And though steam engines are as the angels in heaven, with respect to matrimony, yet in their reproduction of machinery we seem to catch a glimpse of the extraordinary vicarious arrangement whereby it is not impossible that the reproductive system of the mechanical world will be always carried on," noted Samuel Butler in 1865.[63] Seven years later he was more explicit about the reproductive strategies of machines: "Surely if a machine is able to reproduce another machine systematically, we may say that it has a reproductive system. What is a reproductive system, if it be not a system for reproduction? And how few of the machines are there which have not been produced systematically by other machines? ... Each one of ourselves has sprung from minute animalcules whose entity was entirely distinct from our own, and which acted after their kind with no thought or heed of what we might think about it. These little creatures are part of our own reproductive system; then why not we part of that of the machines? ... We are misled by considering any complicated machine as a single thing; in truth it is a city or society, each member of which was bred truly after its kind."[64]

In Butler's time the business of replication was conveyed from generation to generation by engineers. The kingdom of machines might be growing and evolving, but to view machines as organisms was premature. "The kingdoms of living matter and of not-living matter are under one system of laws," declared Butler's adversary Thomas Huxley, "and there is a perfect freedom of exchange and transit from one to the other. But no claim to biological nationality is valid except birth."[65]

Samuel Butler died in 1902. The mechanical kingdom continued to proliferate, spawning a cascade of new species while others, such as steam engines, became extinct. With the advent of electronic digital computers the sense of anticipation—and an interest in Butler's prophecies—was renewed. These machines showed signs of intelligence, and intelligence is a sign of life, even skeptics have agreed. But to ascribe a living intelligence to computers confuses causes with symptoms and was soon shown to be premature.

Computers may turn out to be less important as an end product of technological evolution and more important as catalysts facilitating evolutionary processes through the incubation and propagation of self-replicating filaments of code. As Erasmus Darwin and his Lunar Circle characterized the age that brought mechanical and electromagnetic metabolism to life, so John von Neumann and his circle of engineers and programmers characterized the origins, two centuries later, of self-replicating strings of bits. In 1948, von Neumann delivered his "General and Logical Theory of Automata," from which my father, in his *Origins of Life*, condensed the essential truths that "metabolism and replication, however intricately they may be linked in the biological world as it now exists, are logically separable. It is logically possible to postulate organisms that are composed of pure hardware and capable of metabolism but incapable of replication. It is also possible to postulate organisms that are composed of pure software and capable of replication but incapable of metabolism."[66]

The origins of life as we know it—and life as we are creating it—are to be found in the cross-fertilization between self-sustaining metabolism and self-replicating code. The coalescence of the kingdom of numbers with the kingdom of machines has been incubating for over three hundred years. By the time Erasmus Darwin began experimenting with the effects of electrochemical signals conveyed through his patients' twitching nerves, the essential principles of electromagnetic telecommunications had already been conceived. The results include not only human communications at a distance, and the local replication and preservation of data over time, but human communication *with* machines and, increasingly, communication among machines themselves. To put things in historical perspective, back to Samuel Butler, in New Zealand, in 1863 . . .

The harbor of Port Lyttleton, some seven miles southeast of Christchurch, is formed by the crater of an extinct volcano, surrounded by steep hills. When Samuel Butler arrived in New Zealand in January 1860, communication between the two settlements was either by a rough bridle path overland or around the exposed headlands via sea. The colonists soon connected their two communities via telegraph (the first in New Zealand), thereby conveying notice of arriving vessels, the latest wool prices, and other time-sensitive news. Communication between Christchurch and Lyttleton was no longer delayed by the obstacle of the Port Hills, but by the time consumed by a local echo in the first few feet of the circuit—the telegraph operator's nerves. The telegraph opened on 1 July 1862 and inspired a letter that appeared in the Canterbury *Press* on 15 September 1863. "Why should I write to the newspapers instead of to the machines themselves, why not summon a monster meeting of machines, place the steam engine in the chair, and hold a council of war?" asked the anonymous "mad correspondent." "I answer, the time is not yet ripe for this. . . . Our plan is to turn man's besotted enthusiasm to our own advantage, to make him develop us to the utmost, and find himself enslaved unawares.

"My object is to do my humble share towards pointing out what is the ultimatum, the ne plus ultra of perfection in mechanized development," the writer continued, "even though that end be so far off that only a Darwinian posterity can arrive at it. I therefore venture to suggest that we declare machinery and the general development of the human race to be well and effectually completed when—when—when—Like the woman in white, I had almost committed myself of my secret. Nay, this is telling too much. I must content myself with disclosing something less than the whole. I will give a great step, but not the last. We will say then that a considerable advance has been made in mechanical development, when all men, in all places, without any loss of time, are cognizant through their senses, of all that they desire to be cognizant of in all other places, at a low rate of charge, so that the back country squatter may hear his wool sold in London and deal with the buyer himself—may sit in his own chair in a back country hut and hear the performance of Israel in Ægypt at Exeter Hall—may taste an ice on the Rakaia, which he is paying for and receiving in the Italian opera house Covent garden. Multiply instance *ad libitum*—this is the grand annihilation of time and place which we are all striving for, and which in one small part we have been permitted to see actually realised."[67]

This letter, bearing the stamp of Samuel Butler in style if not in name, was signed "Lunaticus." One hundred years after Erasmus Darwin gathered his circle of Lunaticks in the English Midlands, a strand of telegraph wire was uncoiled at the antipodes of the earth. Sparked by the transit of a few pulses of electromagnetic code over this embryonic fragment of a net, Samuel Butler foresaw the evolution, perhaps not so far off as he imagined, of that phenomenon, somewhere between mechanism and organism, now manifested as the World Wide Web.

Butler was a satirist by trade, a prophet who knew that prophets who take themselves too seriously end up preaching to an audience of one. "There is a period in the evening, or more generally towards the still small hours of the morning, in which we so far unbend as to take a single glass of hot whiskey and water," he admitted to the readers of the Canterbury *Press* in 1865. "We will neither defend the practice nor excuse it. We state it as a fact which must be borne in mind by the readers of this article; for we know not how, whether it be the inspiration of the drink, or the relief from the harassing work with which the day has been occupied, or from whatever other cause, yet we are certainly liable about this time to such a prophetic influence as we seldom else experience."[68]

Although multimedia communication has so far neglected our sense of taste, Butler got the rest of it right. Anticipating modern purveyors of global networking, he presented his vision in terms of the content to which people would subscribe and locally understand. He knew that the development of telecommunications, facilitating the exchange of intelligence among human beings, brings with it the exchange of intelligence among machines. The drift of his thinking and the unspoken secret that he hinted at may be detected in a later comment, in *Unconscious Memory*, that "the component cells of our bodies unite to form our single individuality, of which it is not likely they have a conception, and with which they have probably only the same partial and imperfect sympathy as we, the body corporate, have with them."[69]

3

The General Wind

Far as all such engines must ever be placed at an immeasurable interval below the simplest of Nature's works, yet, from the vastness of those cycles which even human contrivance in some cases unfolds to our view, we may perhaps be enabled to form a faint estimate of the magnitude of that lowest step in the chain of reasoning, which leads us up to Nature's God.

—CHARLES BABBAGE[1]

A t the close of a long and otherwise flattering letter, twenty-four-year-old Gottfried Wilhelm von Leibniz complained to eighty-two-year-old Thomas Hobbes in 1670, "I also wish that you might say something more clearly about the nature of the mind."[2] Ever since Hobbes and Leibniz, the nature of mind has been inextricably linked to the nature of machines. Mind has either been defined as a property *of* the machine, mysterious as the inner workings may be, or, alternatively, as a property *beyond* the machine, no less mysterious through being so diffused. Just as a cathedral organ, no matter how elaborate, cannot produce music without wind, philosophers have sought to identify the invisible ingredient that leads from the predictability of logic to the unpredictability of mind. Can the unlimited power of a mind be evoked by the limited substance of a machine?

Leibniz's lifelong reflections on the nature of mind culminated in his *Monadology* of 1714, a universe of elementary mental particles that he called *monads*, or "little minds." These entelechies (the local actualization of a universal mind) reflect in their own inner state the state of the universe as a whole. According to Leibniz, relation gave rise to substance, not, as Newton had it, the other way around. Our universe had been selected from an infinity of possible universes, explained Leibniz, so that a minimum of laws would lead to a maximum diversity of results. God was the supreme intelligence at both extremes of the scale. As Olaf Stapledon would later put it,

35

"God, who created all things in the beginning, is himself created by all things in the end."[3]

Leibniz enrolled in the University of Leipzig as a law student at age fifteen. His affinity for the law was a mixed blessing, which exercised and supported his interests in formal logic but throughout his life distracted him from scientific work. He became a mathematician by way of reputation, yet remained a courtier by way of life. "Leibniz's tragedy was that he met the lawyers before the scientists," concluded E. T. Bell.[4] Nonetheless, Leibniz made fundamental contributions to mathematics on several fronts. The development of a calculus of continuous functions he shared, controversially, with Isaac Newton, while in combinatorial analysis—the study of relations among discrete sets—Leibniz had the field to himself.

Leibniz continued reasoning about reasoning where Hobbes left off. He attempted to formalize a consistent system of logic, language, and mathematics by means of an alphabet of unambiguous symbols manipulated according to definite rules. A fascination with formal systems and an insight into mechanical computation were combined in the person of Leibniz from the start. Encouraged by his initial steps toward symbolic logic—and by a working model of a calculating machine—Leibniz declared in 1675 to Henry Oldenburg, secretary of the Royal Society and Leibniz's go-between with Isaac Newton, that "the time will come, and come soon, in which we shall have a knowledge of God and mind that is not less certain than that of figures and numbers, and in which the invention of machines will be no more difficult than the construction of problems in geometry."[5]

Leibniz thus helped set in motion the two great movements that led to the age of digital computers in which we live. His calculating machine, demonstrated to the Royal Society of London on 22 January 1673, opened a new era in the mechanization of arithmetic. With his logical calculus, or *calculus ratiocinator,* he took the first steps toward the arithmetization of logic, and in his grand but fragmentary vision of a "universal symbolistic in which all truths of reason would be reduced to a kind of calculus," he predicted the arithmetization of thought itself.[6]

Leibniz credited the invention of his calculator to the inspiration of "an example of the most fortunate genius,"[7] the adding machine constructed in 1642 by nineteen-year-old Blaise Pascal. Leibniz's invention, like Pascal's, was commercially unsuccessful ("It was not made for those who sell oil or sardines," Leibniz explained)[8] and is now represented by a single specimen that was lost in an attic until 1879. But the Leibniz "stepped reckoner," executing a multiple-digit addition cycle with each revolution of a stepped cylindrical gear, was reinvented several times, advancing the mechanization of accounting and finance as inexorably as the mechanization of industry was driven by steam. "Many applications will be found for this machine," wrote Leibniz in 1685, "for it is unworthy of excellent men to lose hours like slaves in the labor of calculation, which could be safely relegated to anyone else if the machine were used."[9]

Leibniz's calculator, based on decimal arithmetic, was widely imitated, whereas his work in binary arithmetic languished for centuries before being embodied in mechanical form. He himself credited the invention of binary notation to the Chinese, seeing in the binary hexagrams of the *I Ching* the remnants of mathematical insights long obscured. "The 64 figures represent a Binary Arithmetic . . . which I have rediscovered some thousands of years later. . . . In Binary Arithmetic there are only two signs, 0 and 1, with which we can write all numbers. . . . I have since found that it further expresses the logic of dichotomies which is of the greatest use."[10] Leibniz saw binary arithmetic as both a practical aid to calculation and a logical calculus leading from the simple to the complex. Multiplication and division could be simplified by switching to numbers encoded in binary form. His notes show the development of simple algorithms, or step-by-step mechanical procedures, for translating between decimal and binary notation and for performing the basic functions of arithmetic as mechanically iterated operations on strings of 0s and 1s.

In 1679, Leibniz imagined a digital computer in which binary numbers were represented by spherical pellets, circulating within a kind of pinball machine governed by a rudimentary form of punched-card control. "This [binary] calculus could be implemented by a machine (without wheels)," he wrote, "in the following manner, easily to be sure and without effort. A container shall be provided with holes in such a way that they can be opened and closed. They are to be open at those places that correspond to a 1 and remain closed at those that correspond to a 0. Through the opened gates small cubes or marbles are to fall into tracks, through the others nothing. It [the gate array] is to be shifted from column to column as required."[11] In the shift registers at the heart of the electronic microprocessor voltage gradients and pulses of electrons have taken the place of gravity and marbles, but otherwise things are running exactly as envisioned by Leibniz in 1679.

Leibniz's ambitions in symbolic logic were similarly prescient, but also incomplete. He believed that "a kind of alphabet of human thoughts can be worked out and that everything can be discovered and judged by a comparison of the letters of this alphabet and an analysis of the words made from them."[12] But he never got around to completing more than a bare outline of his plan. "I think that a few selected men could finish the matter in five years," he claimed, with an optimism echoed by developers of computer operating systems from time to time. "It would take them only two, however, to work out, by an infallible calculus, the doctrines most useful for life, that is, those of morality and metaphysics. . . . Once the characteristic numbers for most concepts have been set up, however, the human race will have a new kind of instrument which will increase the power of the mind much more than optical lenses strengthen the eyes. . . . Reason will be right beyond all doubt only when it is everywhere as clear and certain as only arithmetic has been until now."[13]

Leibniz proposed a universal coding of natural language based on underlying logical relationships and forms. Primary concepts would be represented by prime numbers. From this initial mapping between numbers and ideas a grand, omnipotent combinatorial system could be constructed by arithmetic alone. Leibniz saw that the correspondence between logic and mechanism worked both ways. To his "Studies in a Geometry of Situation," sent to Christiaan Huygens in 1679, Leibniz appended the observation that "one could carry out the description of a machine, no matter how complicated, in characters which would be merely the letters of the alphabet, and so provide the mind with a method of knowing the machine and all its parts."[14]

This ambition was fulfilled, some 150 years later, by the English mathematician, engineer, and patron saint of the programmable computer, Charles Babbage (1791–1871). "By a new system of very simple signs I ultimately succeeded in rendering the most complicated machine capable of explanation almost without the aid of words," wrote Babbage, describing the notation developed in working out the design of his series of difference and analytical engines over the years. "I have called this system of signs the Mechanical Notation. . . . It has given us a new demonstrative science, namely, that of proving that any given machine can or cannot exist."[15]

Babbage's analytical engine aimed to multiply or divide two fifty-digit numbers, to one hundred decimal places, in under a minute's time. Its mechanism was detailed in hundreds of drawings, but only a fragment of it was ever built. The engine could be programmed to evaluate polynomial expressions of unlimited degree, passing intermediate results back and forth between the engine's "store" of internal memory (one thousand registers of fifty decimal digits each) and its arithmetic "mill."

A design of such unprecedented complexity, reported Babbage's associate Harry Wilmot Buxton, "seemed well calculated to overwhelm the most robust intellect. It was therefore only by means of his happy invention of the 'Mechanical Notation,' that he was enabled to alleviate this arduous labour, and partially relieve his brain from a pressure which menaced his bodily health."[16] Although framed in a dialect of gears, levers, and camshafts, Babbage anticipated the formal languages and timing diagrams that brought mechanical logic into the age of relays, vacuum tubes, transistors, microprocessors, and beyond. Computers have been paying their respects to Babbage ever since.

The analytical engine linked the seventeenth-century visions of Hobbes and Leibniz to the twentieth century that digital computation has so transformed. "Mr. Babbage entertained no doubt," wrote Buxton, "of the possibility of extending the powers of the Analytical Engine, far beyond the domain of abstract analysis, and Thomas Hobbes of Malmsbury, as early as 1650, seems to have remarked the analogy existing between the operations of mental computation, and those other operations of the mind."[17] As Hobbes had inspired Leibniz— who admitted that even those works with which he disagreed "usually contain something good and ingenious"[18]—Leibniz in turn inspired Babbage with computational ideas. As an undergraduate at Cambridge, Babbage founded the Analytical Society, seeking to revitalize English mathematics by following the continental lead. To a university enamored of Newton and a nation at war with France, this was a controversial stance. The argument over whether to favor Newton's or Leibniz's notation for the calculus reflected an underlying divergence in mathematical philosophy: Newton's seeking to encompass the kingdom of nature within the mathematical bounds of natural law versus Leibniz's seeking to construct the unbounded kingdom of God from mathematical truth. Babbage believed that the powers of a calculating engine would illuminate both approaches to natural philosophy with the clarity that numbers alone can provide. His motives for the invention of his engines went deeper than the errors that plagued the manually calculated mathematical tables of his time.

Babbage stated that it was "either in 1812 or 1813" that he began "thinking that all these Tables . . . might be calculated by machinery," thereby avoiding mental drudgery as well as the inevitable errors that, especially in tables used for navigation, presented a hazard to life and limb.[19] Although fond of pointing out the *"Erratum* of the *Erratum* of the *Errata* of Taylor's Logarithms" in the *Nautical Almanac* for 1836, Babbage saw the creation of accurate tables as only one of many applications for his machine. Buxton related the genesis of Babbage's idea: "It was whilst endeavouring to reconcile the difficulties involved in the several ideas of Leibnitz and Newton, that Mr. Babbage was led to . . . consider the possibility of making actual motion, under certain conditions, the index of the quantities generated, in arithmetical operations. Thus motion, by means of figure wheels might be . . . conveyed or transferred through racks or other contrivances to successive columns of other wheels, and dealt with arithmetically, under any conditions, which the mechanist thought proper to impose.[20]

A working model of a portion of Babbage's difference engine was soon constructed and successfully used, but completion of a larger engine was bogged down by design changes, engineering difficulties, and negotiations over government support.

Babbage began the design of the analytical engine in 1834 and was still constructing pieces of it in his own workshops when he was eighty years of age. The engine was designed to be able to manipulate its own internal storage registers while reading and writing to and from an unbounded storage medium—strings of punched pasteboard cards, adapted by Babbage from those used by the card-controlled Jacquard loom. A prototype Jacquard mechanism had been introduced in 1801; some eleven thousand Jacquard looms were in use by 1812. In specifying punched-card peripheral equipment, Babbage set a precedent that stood for 150 years. The technology was proven, available, and suited to performing complex functions on extensive data sets. (One demonstration weaving project, a silk portrait of Jacquard, required a sequence of twenty-four thousand cards.) Babbage designated two species of cards for his machine: operation cards, containing programs to be executed; and variable cards, which indexed the location of data in the machine's internal store that was to be processed by the mill. Microprograms were kept at hand in the mill, encoded on toothed cylinders and positioned similarly to the read-only memory (ROM) plugged into the motherboards of most computers today. The analytical engine possessed theoretically unlimited powers of calculation, the recognition of which by Babbage anticipated Alan Turing's demonstration, a hundred years later, that even a very simple analytical engine, given an unlimited supply of cards, can compute any computable function—though it may take a very, very long time.

"The Analytical Engine is therefore a machine of the most general nature," explained Babbage, who understood the value of reusable coding, although programs (referred to as "laws of operation") were not so named. "The Analytical Engine will possess a library of its own. Every set of cards once made will at any future time reproduce the calculations for which it was first arranged."[21] Babbage pursued the design, engineering, and programming of the analytical engine to a stage at which the machine could probably have been built. With extensive debugging, it might have even worked. In 1991, to commemorate the bicentenary of Babbage's birth, a team led by Doron Swade at the Science Museum in London assembled some four thousand components reconstructed according to Babbage's 1847 drawings of Difference Engine No. 2. The three-ton device "flawlessly performed its first major calculation," and "affirmed that Babbage's failures were ones of practical accomplishment, not of design."[22]

Babbage associated with the famous and powerful of his day ("I . . . regularly attended his famous evening parties," recalled Charles Darwin)[23] and held Isaac Newton's Lucasian chair at Cambridge University from 1828 to 1839. His most celebrated collaboration was with the mathematically gifted Lady Ada Augusta Lovelace, daughter of the poet Lord Byron and protégée not only of Babbage but, to a lesser extent, of logician Augustus de Morgan, who was at the same time encouraging work on the *Laws of Thought* by George Boole. Lovelace's extensive notes, appended to her translation of Luigi Menabrea's description of the analytical engine (compiled after Babbage's visit to Italy in 1841 as a guest of the future prime minister) convey the potential she saw in Babbage's machine. "In enabling mechanism to combine together *general* symbols, in successions of unlimited variety and extent, a uniting link is established between the operations of matter and the abstract mental processes of the *most abstract* branch of mathematical science," wrote Lovelace. "A new, a vast, and a powerful language is developed for the future use of analysis, in which to wield its truths. . . . We are not aware of its being on record that anything partaking of the nature of what is so well designated the *Analytical* Engine has been hitherto proposed, or even thought of; as a practical possibility, any more than the idea of a thinking or of a reasoning machine."[24]

Did Babbage grasp the principles of the stored-program digital computer, or has hindsight (and mythology surrounding Lady Lovelace) read too much of the twentieth century into his ideas? Considering the arrangements made for the engine to execute conditionally branched instructions and to change its own course of operation according to a preconceived but not precalculated plan, the evidence in Babbage's favor is substantial. But he never explicitly discussed loading instructions as well as data in the store. In his *Ninth Bridgewater Treatise*, which makes a series of convincing arguments for viewing the universe as a stored-program computer (with God as programmer and miracles as improbable but not impossible subroutines), Babbage related, "I had determined to invest the invention with a degree of generality which should include a wide range of mathematical power; and I was well aware that the mechanical generalisations I had organised contained within them much more than I had leisure to study, and some things which will probably remain unproductive to a far distant day."[25]

Babbage saw digital computers as instruments by which to catalog otherwise inaccessible details of natural religion—the mind of God as revealed by computing the results of his work. He believed that faster, more powerful computers would banish doubt, restore faith, and allow human beings to calculate fragments of incalculable truth. "A time may arrive when, by the progress of knowledge, internal evidence of the truth of revelation may start into existence with all the force that can be derived from the testimony of the senses," he exclaimed.[26]

Babbage was also a prophet of telecommunications. By analyzing the operations of the British postal system, he determined that the cost of conveying letters was governed more by switching than by distance, and he advocated flat-rate postage based on weight. Instituted by Rowland Hill in 1840 as the penny post, Babbage's reforms led to sorting and routing algorithms followed by all subsequent packet-switched information nets. To eliminate the wasted time and energy of forwarding packets of letters by horse, Babbage proposed a mechanically driven communications network that would operate over steel wires three to five miles in length and terminate in nodes where "a man ought to reside in a small station-house." A small metal cylinder containing messages and traveling along the wire "would be conveyed speedily to the next station, where it would be removed by the attendant to the commencement of the next wire, and so forwarded." Babbage knew that it would soon be possible to eliminate the transmission of paper as well as the transmission of the horse. "The stretched wire might itself be available for a species of telegraphic communication yet more rapid," he suggested in 1835.[27]

Babbage was in contact with Joseph Henry and other electrical pioneers but made no attempt to adopt electrical powers in his work. The clock rate of his computer would have been governed by the speed of bronze and iron, with access to its internal memory depending on brute force to shift and spin through an address space with a mass of several tons. But given enough time, enough horsepower, and enough cards, the analytical engine would get the job—any job—done. When Babbage compiled his autobiographical *Passages from the*

Life of a Philosopher in 1864, he concluded that "the whole of the conditions which enable a finite machine to make calculations of unlimited extent are fulfilled. . . . I have converted the infinity of space, which was required by the conditions of the problem, into the infinity of time."[28]

While Babbage was realizing Leibniz's ambitions for the mechanization of arithmetic, Leibniz's agenda for the formalization of mental processes was brought closer to fruition through the late-blooming mathematical career of an English schoolmaster named George Boole (1815–1864). The self-educated son of a Lincoln shopkeeper and boot maker, Boole developed a precise system of logic—Boolean algebra—that has supported the foundations of pure mathematics and computer science ever since. Where Leibniz prophesied the general powers of symbolic logic, Boole extracted a working system from first principles. Intended to provide mathematical foundations for the development of logic, Boolean algebra has also provided logical foundations for new areas of mathematics such as set theory, lattice theory, and topology, a success that was not entirely unforeseen. Boole's initial results were presented in a thin volume, *The Mathematical Analysis of Logic* (1847), followed by *An Investigation of the Laws of Thought, on which are founded the mathematical theories of Logic and Probabilities* (1854).

Boole's goal was "to investigate the fundamental laws of those operations of the mind by which reasoning is performed; to give expression to them in the symbolical language of a Calculus, and . . . to make that method itself the basis of a general method for the application of the mathematical doctrine of Probabilities; and, finally, to collect from the various elements of truth brought to view in the course of these inquiries some probable intimations concerning the nature and constitution of the human mind."[29] Boole's real achievement, however, was the construction of a system of logic rigorous enough to stand on its own as mathematics, independent of the mysteries of mind.

Ordinary algebra uses symbols in place of quantities, allowing systematic analysis of algebraic functions irrespective of actual magnitudes (or what they happen to represent). In Boolean algebra, symbols represent classes of things and Boolean functions the logical relationships between them, allowing the formulation of what are intuitively perceived as concepts or ideas. In reducing logic to its barest essence, Boole's algebra consisted of the symbols $+$, $-$, $\times$, and $=$, representing the logical operations "or," "not," "and," and "identity," operating on variables (x, y, z, etc.) restricted to the values 0 and 1. The

Boolean system, seeded with a minimum of axioms and postulates, assumes as initial conditions only the existence of duality—the distinction between nothing and everything; between true and false; between on and off; between the numbers 0 and 1. Boole's laws were configured so as to correspond not only with ordinary logic but also with binary arithmetic, thereby establishing a bridge between logic and arithmetic that communicates both ways. Using Boolean algebra, logic can be constructed from arithmetic and arithmetic can be constructed from logic. The depth of this functional equivalence, on which the effectiveness of digital computers depends, represents the common ancestry of both mathematics and logic in the genesis of the many from the one.

The success of Boolean algebra has left us with the impression of Boole's *Laws of Thought* as an exact, all-or-nothing system of bivalent logic, as intolerant of error and ambiguity as the integrated circuits and binary coding that have made Boolean logic a household word today. It is something of a historical, technical accident that the logical reliability of the integrated circuit has produced this enduring monument to the precisely true–false Boolean algebra that constitutes the first half of Boole's book, while allowing us to largely ignore the probabilistic and statistical ("fuzzy") logics that made up the final two sections of his work. In the days of vacuum tubes, relays, and hand-soldered plugboards the isomorphism between switching circuits and Boolean algebra was recognized in theory, but in actual practice the function of electrical components over millions of cycles fell short. As Herman Goldstine has pointed out, recalling the ENIAC's seventeen thousand vacuum tubes and one-hundred-kilocycle clock rate, this meant 1.7 billion opportunities per second for a vacuum tube to exhibit logical misbehavior—and occasionally one did.[30] In his last year of life, as one of his final bequests of insight to the successors of the ENIAC, John von Neumann published "Probabilistic Logics and the Synthesis of Reliable Organisms from Unreliable Components,"[31] which is closer to the true spirit of Boole's *Laws of Thought* than is the infallible Boolean logic with which solid-state electronics has surrounded us today.

Boole (and von Neumann) showed how individually indeterminate phenomena could nonetheless be counted on, digitally, to produce logically certain results. "We possess theoretically in all cases, and practically, so far as the requisite labour of calculation may be supplied, the means of evolving from statistical records the seeds of general truths which lie buried amid the mass of figures,"[32] wrote Boole, foreshadowing von Neumann's conclusion that the fundamental "machine" language of a brain constructed from imperfect neurons must be statistical in nature, at a level deeper than the logical processes that appear fundamental to us.

Boole also recognized that error and unpredictability, however foreign to the laws of Newtonian physics and formal logic, may be essential to our ability to think. "The slightest attention to the processes of the intellectual world," concluded Boole, "reveals to us another state of things. The mathematical laws of reasoning are, properly speaking, the laws of right reasoning only, and their actual transgression is a perpetually recurring phenomenon. Error, which has no place in the material system, occupies a large one here."[33]

An unbridged chasm separates our understanding of the logic of mental processes from our understanding of how these processes are executed in the brain. "One finds there only a confused mass in which nothing unusual appears but which nevertheless conceals some kind of filaments of a fineness much greater than that of a spider's web," wrote Leibniz in 1702. "For the subtlety of the spirits contained in these passages is equal to that of light rays themselves."[34]

Among the first to attempt to close the gap between neurology and mind was the English physician Alfred Smee (1818–1877), a prolific investigator whose contributions spanned numerous disciplines, from *The Potato Plant, Its uses and properties, together with the cause of the present malady* (1846) to a pioneering and widely reprinted sixpence broadsheet, *Accidents and Emergencies; A Guide for their Treatment before the arrival of Medical Aid*. The son of William Smee, accountant general to the Bank of England, Alfred grew up within the walled compound of the bank, spending long hours in an improvised laboratory on the ground floor of his father's residence, where he invented a new system for splinting fractures (1839), Smee's battery (1840), and other innovations that gained him fame if not reward. In 1841, at the age of twenty-two, he was appointed surgeon to the Bank of England, "a post which had been especially created for him by the directors ... who thought that the bank could turn his scientific genius to good account."[35] Smee, who had a passion for all things electrical, invented the electrotype plate and by applying the technique to the printing of counterfeit-resistant English banknotes proved the directors' instincts right. Smee's two great interests were combined in a wide-ranging work of electrophysiology, *Elements of Electro-Biology; or, the Voltaic mechanism of Man* (1849), abridged and illustrated for a popular audience under the title *Instinct and Reason* in 1850. Smee introduced the use of electricity in diagnostic medicine and published a pamphlet entitled *The Detection of Needles, and other Steel Instruments, impacted in the Human Frame* (1844)—an occurrence all too common in the industrial workplace of his day.

Smee worked both in theory and in the laboratory to explain the electrochemical basis of vision, sensation, memory, logic, and the origination and recombination of ideas. He believed that the mental powers of animals, human beings, and mechanisms were different not in kind but in degree. His definition of consciousness has seen scant improvement in 150 years. "When an image is produced by an action upon the external senses, the actions on the organs of sense concur with the actions in the brain; and the image is then a *Reality*. When an image occurs to the mind without a corresponding simultaneous action of the body, it is called a *Thought*. The power to distinguish between a thought and a reality is called *Consciousness*," he wrote in his *Principles of the Human Mind deduced from Physical Laws,* published in 1849.[36] As Leibniz envisioned the principles of digital computation, so Alfred Smee envisioned the crude beginnings of a theory of neural nets. "On attending the Physiological Lectures of Professor Mayo, I was remarkably struck with the unsatisfactory account of the functions of the brain, and I was surprised that so little appeared to have been done in connecting mental operations with that organ to which they were due," he wrote in the introduction to his *Process of Thought Adapted to Words and Language, together with a description of the Relational and Differential Machines.*[37]

After considering what little was known concerning neural function at the time, Smee concluded that "every idea, or action on the brain, is ultimately resolvable into an action on a certain combination of nervous fibres, which is definite and determinable, and, regarding the sum total of the nervous fibres, is a positive result over a certain portion only, which has a distinct and clearly defined limit."[38] He was on the right track, but only half right, since he neglected the concept of neural inhibition that is central to the computational and representational powers of neural nets. His system was based on loosely defined analogies between the branching, combinatorial nature of the nervous system and the branching, combinatorial structure of language, logic, and ideas.

Taking the same top-down approach to semantic analysis that would be followed by the artificial intelligence industry in another hundred years, Smee developed a method for parsing natural language by means of a geometric series of symbols ("cyphers") that would render the meaning of any given sentence exact. "This mode of notation may, at first sight, appear more complicated than ordinary language," he wrote, "but if carefully studied, it will be found to afford us an artificial mode of reasoning, which, although immensely inferior to that which is in actual operation by the elaborate machine furnished us by nature, yet as far as it goes, may be conducted by fixed and immutable laws."[39] By analysis of "this most exact form of language," Smee made the leap between mind and mechanism, concluding that "it is apparent that thought is amenable to fixed principles. By taking advantage of a knowledge of these principles it occurred to me that mechanical contrivances might be formed which should obey similar laws, and give those results which some may have considered only obtainable by the operation of the mind itself."[40]

Unlike later proponents of neural and semantic nets, Smee made no grand promises of thinking machines, but merely suggested the development of small-scale logical automata for research. "When the vast extent of a machine sufficiently large to include all words and sequences is considered, we at once observe the absolute impossibility of forming one for practical purposes, inasmuch as it would cover an area exceeding probably all London," he cautioned. "Nevertheless, those lesser machines containing but a few elements, exemplify the principles of their operation, and demonstrate those laws of induction, deduction and relation, the right use of which cannot fail to render our thoughts more accurate, and our language more precise."[41]

Smee understood the inescapable bureaucracy and rigidly enforced assumptions of formal systems. He suggested, with an unspoken nod to the thirteenth-century *Ars Magna* of Ramon Lull, that one of his differential machines "might be beneficially brought into use by those who use fixed and unchangeable creeds; for if they be arranged correctly then any deviation from them would be immediately registered. It must be apparent that such a machine would not estimate the quality of the creed, but only show whether any new creed, or portion of creed, coincided or not with the former creed. For whether the creed inferred a belief in the true God, in Mohammed, in ibises, crocodiles, or saints, in the power of the Virgin, or winking pictures of her, or the qualities of relics, or the virtues of images, or in the parties' own inspiration, the effect would be the same."[42]

"By using the relational and differential machines together," concluded Smee, "from any definite number of premises the correct answer may be obtained, by a process imitating as far as possible, the natural process of thought."[43] But Smee advised his public to "rely upon the abilities which it has pleased Providence to give to them, and not seek assistance from extraneous sources," and made only passing reference to the potentials of electrical logic machines, keeping the prospect of so upstaging nature to himself. "In animal bodies we really have electro-telegraphic communication in the nervous system," he had written in *Instinct and Reason,* juxtaposing micrographic plates of brain tissue with electrotyped illustrations of how he imagined the electrical network to be configured in the brain. He built his own simple electric telegraph, a system "of a somewhat similar character, as it communicates intelligence from one spot to another,"[44] and connected it to a thermometer in his greenhouse so as to transmit an alarm signal when extreme temperatures threatened his collections of exotic plants. In 1849 he suggested "in a remote and imperfect manner" how to construct an artificial ear that would translate sound into electrical signals and expressed "no doubt but that a perfect acoustic telegraph could be made, which shall be acted upon by sounds, and have the power of transmitting them to any distance."[45]

Speculating how vision might be processed by eye and brain, Smee introduced concepts we now know as pixelization, bit mapping, and image compression. He suggested both digital facsimile and analog television at a time when photography was still in its formative years. "From my experiments I believe that it is sufficiently demonstrated that the light falling upon the [optic] nerve determines a voltaic current which passes through the nerves to the brain," he wrote. "From this fact we might make an artificial eye, if we did but take the labour to aggregate a number of tubes communicating with photo-voltaic circuits. . . . Having one nervous element, it is but a repetition to make an eye; and . . . there is no reason why a view of St. Paul's in London should not be carried to Edinburgh through tubes like the nerves which carry the impression to the brain."[46]

But Smee's loyalty was to the vegetable kingdom, not the kingdom of machines. "There is nothing to prevent man from forming an elaborate engine, which should work by change of matter [i.e., electricity] . . . but . . . he must, with the Psalmist, exclaim, 'Such knowledge is too wonderful and excellent for me.' "[47] Smee devoted the rest of his life to horticulture and ecology, publishing a monumental volume, *My Garden; its Plan and Culture together with a general description of its geology, botany, and natural history* (1872), illustrated with thirteen hundred plates. "Its author has endeavoured, so to speak, to catch Nature, animate and inanimate, in a trap of some seven acres and a half, and to chronicle all its everyday features with a sort of Boswellian fidelity," wrote the *Saturday Review.*[48] Babbage died alone, obsessed by the unfulfilled promise of his machines; Smee died at peace, surrounded by a garden full of grandchildren and plants. "Had Smee lived a few years later," wrote D'arcy Power, "he would have made himself a great reputation as an electrical engineer."[49]

Hobbes's ratiocination, Leibniz's *calculus ratiocinator*, Babbage's mechanical notation, Boole's laws of thought, and Smee's conceptual cyphers all attempted to formalize the correspondence among a mechanical system of things, a mathematical system of symbols, and our mental system of thought and ideas. All approaches to formalization from the time of Hobbes until today have been haunted by similar questions: Is the formalization consistent? Is it complete? Does it correspond, in whole or part, to the real world? To the way we think? These questions hinge on the definition of consistency and completeness, available in two different strengths. A formal system is syntactically, or internally, consistent if and only if the system never proves both a statement and its negation and syntactically complete if one or the other is always proved. The system is semantically consistent, under a particular external interpretation, if and only if it proves only true statements and semantically complete if all true statements can be proved.

In 1931, Austrian logician Kurt Gödel (1906–1978) expanded the horizons of mathematics by proving, for both definitions, that no formal system encompassing elementary arithmetic can be at the same time both consistent and complete. Within any sufficiently powerful and noncontradictory system of language, logic, or arithmetic, it is possible to construct true statements that cannot be proved within the boundaries of the system itself.

Gödel achieved this conclusion by a technique now known as Gödel numbering, whereby all expressions within the language of a given formal system are assigned unique identity numbers and thereby forced to obey the manipulations of a strictly arithmetic bureaucracy from which it is impossible to escape. ("Gödel, having grown up in the Austrian *Kaiserreich*, famous for its bureaucracy, must have been familiar with the process," says my mother, no stranger to bureaucracy, being Swiss.) The Gödel numbering, like the characteristic numbering of Leibniz, is based on an alphabet of primes. But Gödel, unlike Leibniz, provided an explicit coding mechanism so that translation between compound expressions and their Gödel numbers remains a two-way street.

"Metamathematical concepts (assertions) thereby become concepts (assertions) about natural numbers or sequences of such, and therefore (at least partially) expressible in the symbolism of the system ... itself," wrote Gödel in the introduction to his proof.[50] By some ingenious twists of logic and number theory, Gödel constructed a formula, the Gödel sentence, "which asserts its own unprovability" even though it can be perceived by reasoning outside the system as being true. The Gödel sentence is loosely equivalent to a self-referential statement that says, "This statement cannot be proved." But saying this with words and saying it with mathematics are two different things. The Gödel numbering enables the formalization of this self-reference by means of a sentence (G) saying, in effect, "The sentence with Gödel number g cannot be proved," where the details of the system are manipulated so that the Gödel number of G is g. G cannot be proved within the specified system and so it is true. Since, assuming consistency, its negation cannot be proved, the Gödel sentence is therefore formally undecidable, rendering the system incomplete. Where Leibniz and his followers had dreamed of a universal coding that would allow the calculation of all truths, Gödel showed that even a system as simple as ordinary arithmetic could never be made complete. Thus Gödel brought Leibniz's dream of a universal, all-encompassing formalization to an end.

This upheaval in the foundations of mathematics, preceded by a similar upheaval in physics, widened our view of the world. The mathematical territory that Gödel expropriated from the stronghold of consistency and proof was distributed to the surrounding mathematical wilderness in the form of intuition and truth. Does a restriction on the powers of formalization curtail the effectiveness of formal systems (or close approximations) functioning within the limitations and inconsistencies of the real world? Physics became no less powerful by discovering that exact knowledge lies beyond any one observer's reach. Arithmetic became no less useful when shown to be formally incomplete. On the contrary, Gödel demonstrated the ability of even simple arithmetic to construct truths that lie beyond the reach of proof.

This distinction between provability and truth, and a parallel distinction between knowledge and intuition, have been exhibited as evidence to support a distinction between the powers of mechanism and those of mind. Gödel's second incompleteness theorem—showing that no formal system can prove its own consistency—has been construed as limiting the ability of mechanical processes to comprehend levels of meaning that are accessible to our minds. The argument over where to draw this distinction has been going on for a long time. Can machines calculate? Can machines think? Can machines become conscious? Can machines have souls? Although Leibniz believed that the process of thought could be arithmetized and that mechanism could perform the requisite arithmetic, he disagreed with the "strong AI" of Hobbes that reduced everything to mechanism, even our own consciousness or the existence (and corporeal mortality) of a soul.

"Whatever is performed in the body of man and of every animal is no less mechanical than what is performed in a watch," wrote Leibniz to Samuel Clarke.[51] But, in the *Monadology*, Leibniz argued that "perception, and that which depends upon it, are inexplicable by mechanical causes," and he presented a thought experiment to support his views: "Supposing that there were a machine whose structure produced thought, sensation, and perception, we could conceive of it as increased in size with the same proportions until one was able to enter into its interior, as he would into a mill. Now, on going into it he would find only pieces working upon one another, but never would he find anything to explain Perception. It is accordingly in the simple substance, and not in the composite nor in a machine that the Perception is to be sought. Furthermore, there is nothing besides perceptions and their changes to be found in the simple substance. And it is in these alone that all the internal activities of the simple substance can consist."[52]

The difference of opinion between Hobbes (mind being a temporary artifact of ordinary matter when suitably arranged) and Leibniz (mind being a fundamental element of the universe, intrinsic to all things but not to be explained by the arrangement of things themselves) has fueled opposing visions over the past three hundred years. Hobbes and Leibniz both believed in the possibility of intelligent machines; it was over the issue of mechanism's license to a soul, not to an intelligence, that the two philosophers diverged.

Hobbes's God was composed of substance; Leibniz's God was composed of mind. Leibniz argued against Hobbesian materialism to the very end; yet one senses that he knew that the case was far from closed. "These gentlemen who strongly debase the idea of God do the same with the idea of the soul," wrote Leibniz to Princess Caroline in 1716. "One of their sect could easily persuade himself into believing that idea of some of the ancient writers . . . according to which souls are born when the machine is organized to receive it, as organ-pipes are adjusted to receive the general wind."[53]

ON COMPUTABLE
NUMBERS

*In attempting to construct such machines we should not be irrever-
ently usurping His power of creating souls, any more than we are in
the procreation of children: rather we are, in either case, instruments
of His will providing mansions for the souls that He creates.*

—ALAN TURING[1]

I n 1936 the English logician Alan Turing (1912–1954) adjusted the
natural numbers to receive the general wind.

Turing's generation grew up in the mathematical shadow of
Göttingen's David Hilbert (1862–1943), whose ambitious program of
formalization set the stage for mathematics between World War I and
World War II. At the International Congress of Mathematicians in
Paris in 1900, Hilbert delivered a list of twenty-three unsolved
problems, prefaced by his conviction that if a proposition could be
articulated within the language of mathematics then either its proof
or its refutation must exist. From the elements of logic and number
theory—the common language at the foundations of mathematics—
the Hilbert school believed that all mathematical truths could be
reached by a sequence of well-defined logical steps. In 1928, Hilbert
again addressed the International Congress of Mathematicians. He
identified three questions by which to determine whether any
finite—or at least finitely describable—set of rules could define a
closed mathematical universe: Can the foundations be proved consistent (so that a statement and its contradiction cannot ever both be
proved)? Can they be proved complete (so that all true statements can
be proved within the system itself)? Does there exist a decision
procedure that, given any statement expressed in the given language,
will always produce either a finite proof of that statement, or else a
definite construction that refutes it, but never both?

Gödel's incompleteness theorems of 1931 brought Hilbert's ambitions to a halt. Where the Hilbert school had hoped to construct one complete system encompassing all mathematical truths, Gödel proved that no single mathematical system sufficient for ordinary arithmetic could establish its own consistency without external help. To capture the richness of mathematics would take a multiplicity of systems, nourished by truth from outside as well as proof from within.

The question—known as the *Entscheidungsproblem,* or decision problem—of whether a precisely mechanical procedure could distinguish provable and disprovable statements within a given system remained unanswered, entangled with fundamental difficulties as to how the intuitive notion of a mechanical procedure should be mathematically defined. Alan Turing was a newly elected fellow of King's College at Cambridge University, working under the guidance of topologist Maxwell H. A. Newman, when the *Entscheidungsproblem* first attracted his attention. Hilbert's challenge aroused an instinct, prevalent in the aftermath of Gödel, that mathematical problems resistant to strictly mechanical procedures could be proved to exist. Turing's strikingly original approach, completed when he was twenty-four, succeeded in formalizing the previously informal correspondence between "mechanical procedure" and "effectively calculable," linking both concepts to the definition of recursive functions introduced by Gödel in 1931. "By what species of madness," asked A. K. Dewdney, "might one have supposed that all three notions would turn out to be the same?"[2]

Turing sought to prove the existence of noncomputable functions, but he had to establish the nature of computability first. A function—in essence a list of questions and their answers—is effectively calculable if it is possible to list all the answers by following a finite set of explicit instructions (an algorithm) that defines exactly what to do from one moment to the next. A computable function is a function whose values can be determined by a mechanical procedure performed by a machine whose behavior can be mathematically predicted from one moment to the next. Effectively calculable and computable appear to be saying the same thing, in a circular sort of way. Proving this equivalence required extending the third leg of the tripod, the concept of recursive functions, to set the whole structure on mathematically solid ground.

Recursive functions are functions that can be defined by the accumulation and strictly regulated substitution of elementary component parts. As multiplication can be reduced to a series of additions, and addition reduced to repeated iterations of the successor function (counting ahead one integer at a time), so can all recursive functions be deconstructed into a finite number of elemental steps. The list of ingredients is short: the existence of 0, the existence of 1, the concept of a successor, the concept of identity, a least number operator, and some clerical substitution rules. Loosely speaking, these elements require no mathematical skills beyond the ability to count. Obviously the ability to compute depends on the ability to count; *proving* that from the ability to count *all* recursive, computable, or effectively calculable functions can be constructed by clerical procedures alone was less obvious. Patience is substituted for intelligence, with consequences both practical and profound.

Rather than reviewing the work of his predecessors and approaching the *Entscheidungsproblem* over established ground, Turing took off from first principles on his own. He began by constructing an imaginary device now known as the Turing machine. Had Turing more diligently followed the work of Alonzo Church or Emil Post, who anticipated his results, his interest in the *Entscheidungsproblem* might have taken a less original form. "It is almost true to say that Turing succeeded in his analysis because he was not familiar with the work of others," commented Turing's colleague Robin Gandy. "Let us praise the uncluttered mind."[3]

Turing arrived at his machine by a process of elimination. He began with the idea of a computer—which in 1936 meant not a calculating machine but a human being, equipped with pencil, paper, explicit instructions, and time to devote to the subject at hand. He then substituted unambiguous components until nothing but a formal description of "computable" was left. Turing's machine thus consisted of a black box (as simple as a typewriter or as complicated as a human being) able to read and write a finite alphabet of symbols to and from a finite but unbounded length of paper tape—and capable of changing its own "m-configuration," or "state of mind."

"We may compare a man in the process of computing a real number to a machine which is only capable of a finite number of conditions . . . which will be called 'm-configurations,' " wrote Turing. "The machine is supplied with a 'tape' (the analogue of paper) running through it, and divided into sections (called 'squares') each capable of bearing a 'symbol.' At any moment there is just one square . . . which is 'in the machine.' . . . However, by altering its m-configuration the machine can effectively remember some of the symbols which it has 'seen.' . . . In some of the configurations in which the scanned square is blank (i.e., bears no symbol) the machine writes down a new symbol on the scanned square; in other configurations it erases the scanned symbol. The machine may also change the square which is being scanned, but only by shifting it one place to right or left. In addition to any of these operations the m-configuration may be changed."[4]

Turing introduced two fundamental assumptions: discreteness of time and discreteness of state of mind. From the point of view of the Turing machine (and all digital computers before and since), time consists of distinct, atomistic moments, one followed by the next like the ticking of a clock, the frames of a motion picture, or the one-by-one succession of the natural numbers. This presumption of discrete sequence allows us to make sense of the world. Logic assumes the sequence of cause and effect; physical law assumes a sequence of observable events; mathematical proof assumes a sequence of discrete, logical steps. In the Turing machine these step-by-step processes are represented by a sequence of discrete symbols encoded on an unlimited supply of tape and by discrete, sequential changes in what Turing called the machine's state of mind. Turing assumed a finite number of possible states. "If we admitted an infinity of states of mind, some of them will be 'arbitrarily close' and will be confused," Turing explained. "The restriction is not one which seriously affects computation, since the use of more complicated states of mind can be avoided by writing more symbols on the tape."[5]

The Turing machine thus embodies the relationship between a finite, if arbitrarily large, sequence of symbols in space and a finite, if arbitrarily large, sequence of events in time. Turing was careful to remove all traces of intelligence. The machine can do nothing more complicated or knowledgeable at any given moment than make a mark, erase a mark, and move the tape one square to the right or to the left. Each step in the relationship between tape and Turing machine is determined by an instruction table (now called a program) listing all possible internal states, all possible external symbols, and, for every possible combination, what to do (write or erase a symbol, move right or left, change internal state) in the event that combination comes up. The Turing machine follows instructions and never makes mistakes. Potentially complicated behavior does not require complicated states of mind. By taking copious notes the Turing machine can function well enough, if at an ever more tedious pace, with as few as two internal states. Behavioral complexity is equivalent whether embodied in complex states of mind (m-configurations) or complex symbols (or strings of simple symbols) encoded on the tape.

Turing's deceptively simple model produced surprising results. He demonstrated the existence of a "Universal Computing Machine," a single machine that can exactly duplicate the behavior of any other computing machine. The universal machine embodies the concept we now know as software—encoding a description of some other machine as a string of symbols, say, 0s and 1s. When executed by the universal machine, this code produces results equivalent to those of the other machine. All Turing machines, and therefore all computable functions, can be encoded by strings of finite length. Since the number of possible machines is countable but the number of possible functions is not, noncomputable functions (and what Turing referred to as "uncomputable numbers") must exist. Turing was even able to construct, by a method similar to Gödel's, functions that could be given a finite description but could not be computed by finite means. The most important of these was the halting function: given the number of a Turing machine and the number of an input tape, it returns either the value 0 or the value 1 depending on whether the computation will ever come to a halt. Turing called the configurations that halt "circular" and the configurations that keep going indefinitely "circle free," and demonstrated that the unsolvability of the halting problem implies the unsolvability of a broad class of similar problems, including the *Entscheidungsproblem.* Contrary to Hilbert's expectations, no mechanical procedure can determine, in a finite number of steps, whether any given mathematical statement is provable or not.

Finally, Turing showed that his definition of computability was equivalent to the effective calculability of Alonzo Church and the general recursiveness of Stephen Kleene—convincing evidence that these seemingly diverse formalizations of an intuitive concept represent a common and inescapable truth. "By a kind of miracle," as Gödel himself later referred to Turing's definition, the concept of computability transcends the formalism in which it is expressed.[6]

For the Turing machine this was both good news and bad. The good news was that in principle all digital computers are equivalent; any machine that can count, take notes, and follow instructions can compute any computable function, given an unlimited supply of scratch paper and an unlimited length of time. Software (coding) can always be substituted for hardware (switching), enabling rapid adaptation through software while the underlying hardware slowly evolves. The bad news was the existence of mathematical functions that no machine can ever compute in any amount of time.

It is surprising that noncomputable functions, which outnumber computable ones, are so hard to find. It is not just that noncomputable functions are difficult to recognize or awkward to define. We either inhabit a largely computable world or have gravitated toward a computable frame of mind. The big questions—Is human intelligence a computable function? Are there algorithms for life?—may never be answered. But computable functions appear to be doing most of the work. "Non-computable functions may be the most common type of function in theory, but in practice they hardly ever come up," Danny Hillis has explained. "In fact, it is difficult to find a well-defined example of a non-computable function that anybody wants to compute. This suggests that there is some deep connection between computability and the physical world and/or the human mind."[7]

"On Computable Numbers" secured a Proctor Fellowship for its author, who went to Princeton University in 1937 to complete his doctoral thesis under Alonzo Church. Princeton had become a leading center for mathematical logic, with Church, von Neumann, and Gödel presiding over a steady flow of distinguished visitors and a growing circle of permanent refugees from abroad. Turing's theoretical device shifted the foundations of mathematics; the implications of general-purpose computers reached far beyond. During his stay in Princeton, Turing grew impatient and set about building parts of a computer with his own hands. "Turing actually designed an electric multiplier and built the first three or four stages to see if it could be made to work," related Malcolm MacPhail, who lent Turing his key to the small machine shop within the Palmer Physics Laboratory, next to the mathematics department at Fine Hall. "He needed relay-operated switches which, not being commercially available at that time, he built himself. . . . And so, he machined and wound the relays; and to our surprise and delight the calculator worked."[8]

Princeton is a quiet central New Jersey town whose attention still lingers on its importance in the Revolutionary War. Princetonians had rehearsed for a role in the digital revolution more than once. Turing's experiments in 1938 were preceded fifty years earlier by the construction of a mechanical binary logic machine by Allan Marquand (1853–1924). "The new machine was constructed in Princeton during the winter of 1881–82," Marquand reported in 1885, and was "made from the wood of a red-cedar post, which once formed part of the enclosure of Princeton's oldest homestead."[9] Marquand was a professor of art history whose foray into mechanical logic attracted the attention of logician Charles Sanders Peirce (1839–1914). In 1886, Peirce wrote to Marquand, "I think you ought to return to the problem, especially as it is by no means hopeless to expect to make a machine for really very difficult mathematical problems. But you would have to proceed step by step. I think electricity would be the best thing to rely on."[10]

Marquand seems to have lost interest in the project, although a search of his manuscripts by Alonzo Church in the early 1950s turned up a pen-and-ink schematic that "is probably the first circuit of an electrical logic-machine."[11] It is not known whether the machine was ever built, but George W. Patterson, who believed that "this is very likely the first design of an electric data processor of any kind," performed a logical and magnetological analysis and was willing "to give the designer the benefit of the doubt" that the machine would have worked, if not exactly as planned.[12]

Marquand's logic machine was the successor to the "logical piano" developed in the 1860s by the English economist and logician William Stanley Jevons. The appearance of Marquand's "vastly more clear-headed contrivance" prompted Peirce to compose a short paper titled "Logical Machines" (1887), in which he considered "precisely how much of the business of thinking a machine could possibly be made to perform, and what part of it must be left for the living mind."[13] Despite the primitive abilities evidenced so far, Peirce advised consideration of "how to pass from such a machine as that to one corresponding to a Jacquard loom."[14] Ill at ease in more academic surroundings, Peirce spent thirty years working for the U.S. Coast Survey, where his duties included working as a (human) computer on the Nautical Almanac and conducting gravitational research. He compiled large sections of the eight-thousand-page *Century Dictionary* and was both physically and mentally ambidextrous, able to write out a question and its answer using both hands at the same time.

"Every machine is a reasoning machine, in so much as there are certain relations between its parts, which relations involve other relations that were not expressly intended," observed Peirce.[15] Independently of Babbage and ahead of Turing, he considered the implications of more advanced reasoning by machines. "The machine would be utterly devoid of original initiative, and would only do the special kind of thing it had been calculated to do. This, however, is no defect in a machine; we do not want it to do its own business, but ours. . . . We no more want an original machine, than a house-builder would want an original journeyman, or an American board of college trustees would hire an original professor." (Peirce had been subjected to this judgment firsthand.) "The logical machines that have thus far been devised can deal with but a limited number of different letters," continued Peirce. "The unaided mind is also limited in this as in other respects; but, the mind working with a pencil and plenty of paper has no such limitation . . . whatever limits can be assigned to its capacity to-day, may be over-stepped to-morrow."[16] Peirce recognized the power of unbounded storage—the principle underlying Babbage's analytical engine and later formalized by Turing's universal machine.

Which came first, Turing machine or digital computer? It depends on whether precedence is assigned to the chicken or to the egg. Turing's analysis transcended architectural and genealogical particulars to reveal a universal fellowship among digital machines. At the time of "On Computable Numbers," large numbers of Turing machines already existed in the form of punched-card tabulating, calculating, and data-processing machines. These devices mimicked their theoretical archetype by reading a mark on a piece of paper, shifting internal state accordingly, and making another mark somewhere else. Punched-card machines formed extensive systems whose components differentiated into the essential functions (input, output, storage, and central processing) that would characterize the vital organs of all computers in the years ahead.

The punched-card information-processing industry was developed by Herman Hollerith (1860–1929), employed as a special agent for the tenth U.S. census, in 1879. The 1880 census took almost seven years to completely count. If methods were not improved, the 1890 census would not be completed by the time the 1900 census began. Hollerith's supervisor, Dr. John S. Billings, encouraged his protégé to tabulate data by means of perforated cards, citing the precedent of railway tickets but not Babbage's engine or the Jacquard loom. Once a card was punched the data could then be read, sorted, and tabulated by machine. As a demonstration project Billings arranged for Hollerith to tabulate vital statistics for the Baltimore Department of Health. Hollerith made the most of this opportunity, although, as his mother-in-law wrote in 1889, "he is completely tired out. He has been punching cards at the rate of 1,000 per day—and each card has at least a dozen holes. He has done it all with a hand punch and his arm was aching and paining dreadfully. He really looked quite badly."[17] But the system worked.

Hollerith won the contract to tabulate the eleventh U.S. census of 1890, enumerating sixty-two million people using some fifty-six million cards. There were 288 punch positions, storing the equivalent of up to thirty-six 8-bit bytes of information per card. Results more detailed than those of any previous census were completed within two years. Punched-card equipment proliferated relentlessly as techniques developed to meet the ten-year cycle of the census were adapted to other purposes in between. Hollerith incorporated the Tabulating Machine Company in 1896, which was consolidated into the Computing-Tabulating-Recording Company (CTR) in 1911, and then renamed International Business Machines, or IBM, in 1924. Punched cards and perforated tapes not only served to convey and process information, but began to exercise control. In a 1922 *Scientific American* article subtitled "How Strips of Paper Can Endow Inanimate Machines with Brains of Their Own," Emmanuel Scheyer predicted that "in some uncanny way, things will seem to be running themselves."[18]

By the time Turing awakened us to their powers, the age of discrete-state machines was well under way. "There is a great deal more arithmetic and better arithmetic in the world than there used to be," reported Vannevar Bush in October 1936. "This is indicated by the fact that 10,000 tons of cards are used per year, a total of four billion cards.... The end of the development is not in sight."[19] Each card was functionally equivalent to one or more cells of a Turing machine tape. A machine might be programmed to scan the whole card at once (a pattern of holes punched among ten rows and eighty columns representing an alphabet of 2^{800} possible symbols); a single location on the card (the presence or absence of a hole representing an alphabet of 2 possible symbols); or some intermediate configuration that divided the data into fields. It was characteristic of the data-processing industry at that time (and closer to the original concept of a Turing machine than to the state of data processing today) that most of the complexity was represented by the tape (or card sequence) itself, rather than by the machine's internal state of mind. By repeated sorting and other iterated functions, primitive punched-card machines could perform complex operations, but, like the original Turing machine, they had only a small number of possible states.

The fundamental unit of information was the bit; its explicit definition as the contraction of "binary digit" was first noted in an internal Bell Laboratories memo written by John W. Tukey on 9 January 1947,[20] and first published in Claude Shannon's *Mathematical Theory of Communication* in 1948.[21] Shannon's definition was foreshadowed by Vannevar Bush's analysis, in 1936, of the number of "bits of information" that could be stored on a punched card. In those days bits were assigned only fleetingly to electrical or electronic form. Most bits, most of the time, were bits of paper (or bits of missing paper, represented by the chad that was carted off to landfills by the ton). "There is still a great deal of carrying cards from one machine to another, and each problem is unique," wrote Bush, best known for his differential analyzer, the granddaddy of analog computers, and less known for his prediction, in the 1930s, of an inevitable shift toward digital machines. "A master control is here conceivable. This process should also be automatic and performed entirely by machine. It should be sufficient, having punched a stack of cards, then to punch a master card, dictating with complete flexibility the operations to be performed. . . . Such an arrangement would no doubt be soon worked out if there were sufficient commercial demand. . . . Quite a lot can be accomplished by arithmetic alone, if its operations can be performed rapidly enough, and combined. . . . It is time that a numerical machine were built for which the sequence of operations might be varied at will to cover a large field of utility, but just as fully automatic once the sequence is assigned."[22]

Hollerith equipment was used throughout business, industry, and science to tabulate, store, sort, and condense large amounts of disordered information into a more ordered state. Punched-card routines assisted in searching for underlying patterns and analyzing accumulated results. By discovering correspondences between computable functions and incoming data the machines could be used to model relationships and even predict future sequences of events.

A Turing machine can also be configured to run the other way: instead of finding a pattern in an incoming sequence and producing a comprehensible result, it can transform an understandable message into an arbitrarily complicated sequence, producing an incomprehensible result. If scrambled by a computable function (and thereby encoded as a computable number) someone with knowledge of that function can reverse the process to reconstruct the original message at the other end. At the close of World War I, such a cryptographic machine was invented by the German electrical engineer Arthur Scherbius (1878–1929), who proposed it to the German navy, an offer that was declined. Scherbius then founded the Chiffriermaschinen Aktien-Gesellschaft to manufacture the machine, christened the Enigma, for enciphering commercial communications, such as transfers between banks. The machine attracted a modest following, but sales were limited until the German navy changed its mind. Modified versions of the Enigma machine were adopted by the German navy in 1926, the German army in 1928, and the German air force in 1935.

The heart of the Enigma was a series of flat, wheel-shaped rotors, with twenty-six electrical contacts, one for each letter of the alphabet, arranged in a circle on each face. The contacts were connected in unpredictable order so that a signal going in one side of the rotor as a given letter emerged on the other side as something else. There were thus 26! (or 403,291,461,126,605,635,584,000,000) possible wirings for each rotor. Each station in a particular banking or communication network had an assortment of different rotors in matching sets. The Enigma also had a keyboard like a typewriter. Each key closed a circuit that sent a battery current through a stack of three adjacent rotors; the signal returned via a fourth, reflecting rotor (capable of only 7,905,853,580,025 states) that continued the circuit back through the first three rotors in reverse, ending at one of twenty-six light bulbs, which indicated the letter to be used for the enciphered text. The rotors were mechanically coupled to the keyboard like the wheels of an odometer, so that the machine's state of mind changed with every step. But if the recipient had an identical machine, with the same rotors placed in the same positions, the function could be executed in reverse, producing deciphered text.

In September 1939, at the outbreak of World War II, Alan Turing found himself face-to-face with the Enigma, and later encountered its digital successors, known to the British code breakers as Fish. Turing and a rapidly expanding circle of mathematicians, linguists, engineers, technicians, clerks, and chess players, assisted by an indispensable corps of Wrens (Women's Royal Navy Service), were sequestered at a Buckinghamshire estate known as Bletchley Park for the duration of World War II. As guests of the Foreign Office's Government Code and Cypher School the cryptanalysts kept a low profile, although it was difficult to conceal that so many gifted mathematicians (especially chess players) had suddenly dropped out of sight. Suspicious, but not suspicious enough, the German authorities modified the commercial Enigma machine and frequently changed the keys, suspecting internal spies whenever there was evidence of a leak. For more secure communications, especially with the U-boat fleet, an additional rotor position was added as well as an auxiliary plugboard that further scrambled ten pairs of letters, leaving only six letters unchanged. "Thus, the number of possible initial states of the machine at the beginning of the message was about 9×10^{20}. For the U-boats it was about 10^{23}," recalled Irving J. (Jack) Good, who signed on as Turing's statistical assistant, at the age of twenty-five, in May 1941.[23]

For the three-rotor Enigma a brute-force trial-and-error approach would have to test about a thousand states per second to run through all possible configurations in the three billion years since life appeared on earth. A brute-force approach to the four-rotor Enigma would have to test about 200,000 states per second to be assured of a solution in the fifteen billion years since the known universe began. During a critical period of the war, Bletchley Park succeeded in deciphering a strategically significant fraction of intercepted Enigma traffic within a few days or sometimes hours before the intelligence grew stale. This success was a product of human ingenuity on the part of the British matched by human error on the other side. "We benefited greatly from a combination of Nazi bombast and German methodicalness," recalled Peter Hilton, an Oxford undergraduate recruited in January 1942. "Nazi conceit dictated that great military successes should be announced to every German military unit everywhere; and the passion of the German military mind for good order and discipline dictated that these announcements should be made in exactly the same words and sent out at exactly the same time over all channels."[24]

Polish cryptographers provided a head start by decoding three-rotor Enigma messages before the outbreak of the war. Three young Polish mathematicians (Henryk Zygalski, Jerzy Różycki, and Marian Rejewski), assisted by French intelligence and an interest in the German Enigma dating back to an interception by Polish customs officers in 1928, applied ingenious logic to narrow the search for rotor configurations so that electromechanical devices (called "bombas" by the Poles and "bombes" by the British) could apply trial and error to certain subsets that remained. The bombe incremented itself through a space of possibilities, and when a designated clue turned up it came to a stop. (The characteristic ticking, followed by sudden silence, may have given the machine its name.)

With wartime improvements to the Enigma and increasingly frequent rotor changes, even a growing array of far more powerful British bombes, designed with Turing's assistance and mass-produced by the British Tabulating Machine Company, could barely keep up. By the end of 1943, ninety thousand Enigma messages a month were being decrypted, with round-the-clock shifts of cryptanalysts at Bletchley supported by satellite bombe installations at Wavendon, Gayhurst, Stanmore, and Eastcote. "The bombes were bronze-coloured cabinets about eight feet tall and seven feet wide . . . [and] made a considerable noise as the drums revolved, each row at a different speed, so there was not much talking during the eight-hour spell," recalled Diana Payne, who set up (programmed) bombes according to the day's cryptanalytic menus for more than three years. "For technical reasons which I never understood, the bombe would suddenly stop and we took a reading from the drums . . . it was a thrill when the winning stop came from one's own machine."[25]

Fish traffic—longer messages, transmitted automatically in binary code over high-speed cable and radiotelegraph links—presented a challenge beyond the reach of the bombes. Electronic data processing offered the only hope of catching up. A series of punched-tape machines known as Heath Robinson (code-named after an English cartoonist, in the style of Rube Goldberg, "well known for his drawings of ludicrous tasks performed by fantastic machines")[26] were built, on the principle that by simultaneous scanning of two different (and relatively prime) lengths of coded tape as continuous loops, all possible combinations of the two sequences could be compared. Based on standard teleprinter tape and standard 5-bit (Baudot) teleprinter code, but running at high speed through photoelectric heads, the Heath Robinsons used electronic circuits to count, combine, and compare the two sequences by means of Boolean operations performed at a tremendous pace. But it was difficult to maintain synchronization between two tapes. It was then proposed by Thomas H. Flowers, an engineer working for the British Post Office's telecommunications research station at Dollis Hill, to eliminate one of the tapes by transferring its sequence to the internal memory (or state of mind, in Turing's language) of an electronically more complicated but mechanically simpler machine. The internal sequence could then be precisely synchronized to the sequence of pulses input by the tape, which could be run without sprockets at much higher speeds by friction drive.

"The tapes were read at 5,000 characters per second," recalled Jack Good. "There were parallel circuits, so that 25,000 binary digits were handled every second. . . . Teleprinter tapes have 10 characters to the inch, so that the speed of 5,000 characters per second implies a tape speed of nearly 30 miles per hour. I regard the fact that paper teleprinter tape could be run at this speed as one of the great secrets of World War II!"[27] With practice, it was possible to run loops of tape as much as two hundred feet in length, although there were problems with the edges of the tape sawing through stainless-steel guide pins on longer runs. The new machine, constructed under the supervision of Thomas Flowers at the Dollis Hill research station and operated and programmed under the direction of M. H. A. Newman (under whose supervision Turing had written his paper on computable numbers in 1936), was code-named Colossus and incorporated fifteen hundred vacuum tubes, or, as the British more accurately described them, valves. The machine was so successful (and subspecies of Fish so prolific) that by the end of the war ten Colossi were in use, the later versions using twenty-four hundred vacuum tubes. The heaters were never turned off, since reheating was the most likely occasion for tubes to fail. "Ah, the warmth at two A.M. on a damp, cold, English winter!"[28] recalled Howard Campaigne, a U.S. Navy cryptanalyst assigned to Bletchley Park in 1942. By the end of the war the Germans had begun to change the wheel patterns of both Enigma and Fish once a day instead of once a month.

The Fish were of two families: the *Geheimschreiber,* manufactured by Siemens, and the *Schlüsselzusatz,* manufactured by Lorenz. The latter was targeted by the Colossus and known as Tunny to the British, with various subspecies (Jellyfish, Bream, Gurnard, Sturgeon, etc.) representing different branches of German command. The Fish were substantial pieces of automatic teletypewriter equipment that produced a sequence of 0s and 1s (the key) that was then added to the binary representation of an unenciphered (plaintext) message and output for transmission as ordinary 5-bit teletypewriter tape. The machine's twelve code wheels, of unequal length, were circumscribed by a combined total of 501 pins that could be shifted between two positions, giving the system a formidable number (2^{501} or about 10^{150}) of possible states and a period of 1.6×10^{19} digits before the key produced by any given configuration began to repeat. The key was added modulo 2 to the plaintext message (counting by two the way we count hours by twelve, so that $0 + 1 = 1$ and $1 + 1 = 0$), with 1 and 0 represented by the presence or absence of a hole in the tape. Adding the key to the enciphered text a second time would return the original text. Each Fish was a species of Turing machine, and the process by which the Colossi were used to break the various species of Fish was a textbook example of the process by which the function (or partial function) of one Turing machine could be encoded as a subsidiary function of another Turing machine to produce simulated results. The problem, of course, was that the British didn't know the constantly changing state of the Fish; they had to guess.

Colossus was programmed, in Boolean logic mode, by a plug-board and toggle switches at the back of the machine. "The flexible nature of the programming was probably proposed by Newman and perhaps also Turing, both of whom were familiar with Boolean logic, and this flexibility paid off handsomely," recalled I. J. Good. "The mode of operation was for a cryptanalyst to sit at Colossus and issue instructions to a Wren for revised plugging, depending on what was printed on the automatic typewriter. At this stage there was a close synergy between man, woman, and machine, a synergy that was not typical during the next decade of large-scale computers."[29] Colossus did not directly reveal a plaintext message, but, when successful, a succession of clues as to the configuration and initial position of the wheels that had produced the key sequence in use at the time. The search for clues, often assisted by a "crib," or probable string of text, relied on certain subtle statistical characteristics of the German language, a process that remained one of the more closely guarded secrets of the war. In a demonstration of the machine intelligence that would absorb Turing and several of his colleagues in the aftermath of World War II, Colossus was trained to sense the direction of extremely faint thermoclines that distinguished enciphered German from flat-random alphabetic noise. Said Andrew Hodges, "the line between the 'mechanical' and the 'intelligent' was very, very slightly blurred."[30]

Colossus was not a stored-program computer (executing and modifying internally stored instructions), but it came almost as close as the U.S. Army–sponsored ENIAC, and some two years in advance. It was distinguished from other electronic calculators in that it was designed for performing Boolean operations, not producing numerical results. This counted against it by the standards of its day, but as a step toward the modern computer, these logical abilities placed it far ahead.

Turing's role in the history of Colossus remains shrouded by the layers of secrecy that surrounded the project, further obscured by the legendary aura surrounding the universal machine. Good wrote that Turing "made important statistical contributions, but had little to do with the Colossus,"[31] a view supported by Newman, Flowers, and others, although Brian Randell, after extensive interviews with these participants, noted "virtually all the people I have interviewed recollect wartime discussions of his idea of a universal automaton."[32] Peter Hilton wrote that Turing "was, in fact,—and quite consciously and deliberately—inventing the computer as he designed first the 'Bombe' and then the 'Colossus.' "[33] By the time of the actual construction and operation of the Colossi, Turing had moved on to the problem of real-time voice encryption, among other things. Bletchley Park had grown into an operation employing seven thousand people, ten Colossi, innumerable bombes, large arrays of Hollerith equipment, and extensive telecommunications support. The Colossi were among the first programmable, if specialized, electronic digital computers. As an integrated data-processing installation the whole operation was years, if not decades, ahead of its time.

With the end of the war, the computational torch passed to the Americans, even though it was the alumni of Bletchley Park who were first to demonstrate a working stored-program computer (the Manchester Baby Mark I, which ran its first program on 21 June 1948) and first to construct a fully electronic memory (the electrostatic Williams tube). The driving force behind computer development was no longer the logical puzzle of cryptanalysis, but the numerical horsepower required to design atomic bombs. When Bletchley Park disbanded, the Official Secrets Act handicapped those who could not refer openly to their wartime work. The existence of Colossus would not be officially acknowledged for thirty-two years.

That Turing was thinking seriously about computers during the war is best evidenced by his report, produced in the final three months of 1945 for the National Physical Laboratory (NPL), entitled "Proposal for the Development in the Mathematics Division of an Automatic Computing Engine (ACE)."[34] Turing's design was commissioned by J. R. Womersly, superintendent of the Mathematics Division, who became interested in Turing machines before the war and had even suggested building one before strategic priorities intervened. At the end of the war Womersly had been sent to the United States to survey the latest (and still secret) computer developments, including the Harvard Mark I tape-controlled electronic calculator, which he described as "Turing in hardware" in a letter home.[35] Womersly reported to Douglas R. Hartree, who reported to Sir Charles Darwin, Director of NPL and grandson of *the* Charles Darwin. But Darwin was slow to take an interest in Turing's project, and the lumbering pace of the bureaucracy he commanded had already crippled the proposal by the time he applied his influence in an attempt to gain the project full support. Shuffled among a succession of departments, the original proposal was reconsidered to death. Turing's automatic computing engine, like Babbage's analytical engine, was never built.

Turing's proposal "synthesized the concepts of a stored-program universal computer, a floating-point subroutine library, artificial intelligence, details such as a hardware bootstrap loader, and much else."[36] At a time when no such machines were in existence and the von Neumann architecture had only just been proposed, Turing produced a complete description of a million-cycle-per-second computer that foreshadowed the RISC (Reduced Instruction Set Computer) architecture that has now gained prominence after fifty years. The report was accompanied by circuit diagrams, a detailed physical and logical analysis of the internal storage system, sample programs, detailed (if bug-ridden) subroutines, and even an estimated (if unrealistic) cost of £11,200. As Sara Turing later explained, her son's goal was "to see his logical theory of a universal machine, previously set out in his paper 'Computable Numbers,' take concrete form."[37]

Turing's design relied on mercury-filled acoustic delay lines for high-speed storage, a technique developed for processing radar signals by comparing a series of echoes to distinguish things that had moved and later applied to an early generation of computers, although "its programming," as M. H. A. Newman said, "was like catching mice just as they were entering a hole in the wall."[38] A series of electrical pulses, about a microsecond apart, were converted to a train of sound waves circulating in a long tube of mercury equipped with crystal transducers at both ends. About a thousand digits could be stored in the millisecond it took a train of pulses to travel the length of a five-foot "tank." Viewed as part of a finite-state Turing machine, the delay line represented a continuous loop of tape, a thousand squares in length and making a thousand complete passes per second under the read–write head. Turing specified some two hundred tubes, each storing thirty-two words of 32 bits, for a total, "comparable with the memory capacity of a minnow," of about 200,000 bits.[39]

"The property of being digital," announced Turing to the London Mathematical Society in a 1947 lecture on his design, "should be of more interest than that of being electronic."[40] Whether memory took the form of paper tape, vacuum-tube flip-flops, mercury pulse trains, or even papyrus scrolls did not matter as long as discrete symbols could be freely read, written, relocated, and, when so instructed, erased. The concept of random-access memory and the resulting ability to store and manipulate both instructions and data in common is considered to have been the key innovation in the development of electronic digital computers (producing twenty thousand pages of transcripts in the Honeywell–Sperry-Rand patent dispute alone). Both these developments were implicit in the concept of a one-tape Turing machine introduced in 1936. It made no difference whether binary digits (instructions, data, or temporary notes) were stored as sound waves in a vibrating column of mercury or as symbols on paper tape. But the five-channel tape readers of Colossus would have to be run at twelve hundred miles per hour to keep up with a single mercury delay-line store.

Turing's vision for the ACE became bogged down in an institutional quagmire and failed to get off the ground. The routine miracles of war, when Cambridge theoreticians were granted unlimited engineering resources and even the post office could be counted on to deliver new hardware overnight, did not survive the peace. Turing's decision that construction should be contracted out, as had been done for the Colossus, was in hindsight a mistake. But hindsight has shown that his design principles were sound. In May 1950 a partial prototype (the Pilot ACE) was finally built and "proved to be a far more powerful computer than we had expected," wrote J. H. Wilkinson, even though its mercury delay lines only held three hundred words of 32 bits each. "Oddly enough much of its effectiveness sprang from what appeared to be weaknesses resulting from the economy in equipment that dictated its design."[41]

In July 1947, Turing took a leave of absence from NPL, returning to his King's College fellowship for a year. He resigned from NPL in May 1948, accepting an appointment to Manchester University, where M. H. A. Newman was germinating a mathematical computing department with talent from Bletchley Park. Turing, restless as ever, helped get machines and programs up and running at Manchester while his attention wandered to other things. Foremost was his mathematical theory of morphogenesis, which he worked at simulating digitally, writing programs longhand in machine language using his own base-32 notation (the digits reversed to match the patterns of bits as displayed by the Williams-tube store). Another focus was a series of reflections on artificial intelligence, labeled "mechanical intelligence" in language that remains more precise. Here more than ever his iconoclasm found free reign. "An unwillingness to admit the possibility that mankind can have any rivals in intellectual power," Turing wrote in 1948, "occurs as much amongst intellectual people as amongst others: they have more to lose."[42]

Turing's thoughts about hardware and software ranged far ahead of anything in existence at the time. His approach to the question of machine intelligence was as uncluttered as his approach to computable numbers ten years before. He faced the question of incompleteness once again. A brisk trade would soon develop around the rehashing of Gödel's proof of the incompleteness of formal systems, arguing whether this limitation constrained the abilities of computers to duplicate the intelligence and creativity of the human mind. Turing neatly summarized the essence (and weakness) of this convoluted argument in 1947, saying that "in other words then, if a machine is expected to be infallible, it cannot also be intelligent."[43] To Turing this demonstrated not a theoretical obstacle, but simply the need to develop fallible machines able to learn from their own mistakes.

"The argument from Gödel's and other theorems rests essentially on the condition that the machine must not make mistakes," he explained in a sabbatical report submitted to NPL in 1948. "But this is not a requirement for intelligence."[44] Turing made several concrete proposals. He suggested incorporating a random element to create what he referred to as a "learning machine." This proposal avoided the problem of having to specify all possible contingencies in advance by granting the computer an ability to take a wild guess and then either reinforce or discard the guess according to the consequent results. Guesses might be extended not only to external questions, but to modifications in the computer's own instructions. A machine could then learn to teach itself. "What we want is a machine that can learn from experience," wrote Turing. "The possibility of letting the machine alter its own instructions provides the mechanism for this."[45] In 1949, while developing the Manchester Mark I (commissioned by Ferranti Ltd. as the prototype of the first electronic digital computer to be commercially produced), Turing designed a random-number generator that instead of producing pseudorandom numbers by a numerical process, included a source of truly random electronic noise.

Carrying these ideas one step further (although pointing out that "paper interference" with a universal machine was equivalent to "screwdriver interference" with actual parts), Turing developed the concept of "unorganized Machines ... which are largely random in their construction [and] made up from a rather large number N of similar units."[46] He considered a simple model with units capable of two possible states connected by two inputs and one output each, concluding that "machines of this character can behave in a very complicated manner when the number of units is large." Turing showed how such unorganized machines ("about the simplest model of a nervous system") could be made self-modifying and, with proper upbringing, could become more complicated than anything that could be otherwise engineered. The human brain must start out as such an unorganized machine, since only in this way could something so complicated be reproduced.

Turing perceived a parallel between intelligence and "the genetical or evolutionary search by which a combination of genes is looked for, the criterion being survival value. The remarkable success of this search confirms to some extent the idea that intellectual activity consists mainly of various kinds of search."[47] He saw evolutionary computation as the best approach to truly intelligent machines. "Instead of trying to produce a programme to simulate the adult mind, why not rather try to produce one which simulates the child's?" he asked.[48] "Bit by bit one would be able to allow the machine to make more and more 'choices' or 'decisions.' One would eventually find it possible to program it so as to make its behaviour the result of a comparatively small number of general principles. When these became sufficiently general, interference would no longer be necessary, and the machine would have 'grown up.' "[49]

An incremental, trial-and-error path toward artificial intelligence lay ahead. It is a misconception, based on the stereotype of a Turing machine as executing a prearranged program one step at a time, to assume that Turing believed that any single, explicitly programmed serial process would ever capture human intelligence in mechanical form. Turing knew how many interconnected neurons it took to make a brain, and he knew how many brains it took to form a society that could kindle the spark of language and intelligence into flame. He himself had drawn the curtains on Leibniz's illusion of an ideal, completely formalized logical system in 1936. And in 1939 even his own attempt to transcend Gödelian incompleteness by his "Systems of Logic Based on Ordinals" had failed. In this sequel to "On Computable Numbers," prepared in Princeton as his doctoral thesis under Alonzo Church, Turing explored "how far it is possible to eliminate intuition, and leave only ingenuity," noting that since ingenuity can always be replaced by patience, "we do not mind how much ingenuity is required, and therefore assume it to be available in unlimited supply."[50]

Intelligence would never be clean and perfectly organized, but like the brain would remain slippery and disordered in its details. The secret of large, reliable, and flexible machines, as Turing noted, is to construct them, or let them construct themselves, from large numbers of individual parts—independently free to make mistakes, search randomly, and generally act unpredictably so that at a much higher level of the hierarchy the machine appears to be making an intelligent choice. It is an appealing model—advocated by Oliver Selfridge in his *Pandemonium* of 1959, I. J. Good in his *Speculations Concerning the First Ultraintelligent Machine* (1965), and Marvin Minsky in his *Society of Mind* (1985). A similar principle of distributed intelligence (enforced by need-to-know security rules) led to successful code breaking at Bletchley Park.

The Turing machine, as a universal representation of the relations between patterns in space and sequences in time, has given these intuitive models of intelligence a common language that translates freely between concrete and theoretical domains. Turing's machine has grown progressively more universal for sixty years. From McCulloch and Pitts's demonstration of the equivalence between Turing machines and neural nets in 1943 to John von Neumann's statement that "as far as the machine is concerned, let the whole outside world consist of a long paper tape,"[51] the Turing machine has established the measure by which all models of computation have been defined. Only in theories of quantum computation—in which quantum superposition allows multiple states to exist at the same time—have the powers of the discrete-state Turing machine been left behind.

All intelligence is collective. The truth that escaped Leibniz, but captured Turing, is that this intelligence—whether that of a billion neurons, a billion microprocessors, or a billion molecules forming a single cell—arises not from the unfolding of a predetermined master plan, but by the accumulation of random bits of wisdom through the power of small mistakes. The logicians of Bletchley Park breathed the spark of intelligence into the Colossus not by training the machine to recognize the one key that held the answer, but by training it to eliminate the billions of billions of keys that probably wouldn't fit.

Turing broke the mystery of intelligence into bits, but in so doing revealed a greater mystery: how to reconcile the mechanism of intelligence with the unpredictability of mind. The upheaval in logic of the 1930s was reminiscent of the revolution in physics that revealed the certainties of Newtonian mechanics to be uncertainties in disguise. The great mysteries were shifted from the very large to the very small. By means of Turing's machine, all computable processes could be decomposed into elemental steps—just as all mechanical devices can be decomposed into smaller and smaller parts. Leibniz's thought experiment, in which he imagined entering into a thinking machine as into a mill, was embodied by Turing in rigorous form. The mystery of intelligence was replaced by a succession of smaller mysteries—until there lingered only the mystery of mind.

As Leibniz argued that "there must be in the simple substance a plurality of conditions and relations, even though it has no parts,"[52] so Turing's analysis suggests that the powers of mind derive not only from the realm of very large numbers (by combinatorial processes alone) but from the realm of the very small (by the element of chance adhering to any observable event). Hobbes and Leibniz could *both* be right.

5

THE PROVING GROUND

I am thinking about something much more important than bombs. I am thinking about computers.

—JOHN VON NEUMANN[1]

As the human intellect grew sharper in exercising the split-second timing associated with throwing stones at advancing enemies or fleeing prey, so the development of computers was nurtured by problems in ballistics—the science of throwing things at distant targets through the air or, more recently, through space. The close relationship between mathematics and ballistics goes back to Archimedes, Leonardo da Vinci, Galileo, and Isaac Newton, whose apocryphal apple remains the most famous example of an insight into ballistics advancing science as a whole. It was Robert Boyle, in *The Usefulnesse of Mechanical Disciplines to Natural Philosophy,* who introduced the term *balisticks* into the English language in 1671. Boyle classified ballistics as one of the "fatal arts." The precise use of gunpowder was regarded as a humanitarian advance over indiscriminate mayhem, greeted with the zeal for "smart" weapons that continues to this day.

Alan Turing and his colleagues at Bletchley Park were chess players at heart, pitting their combined intelligence against Hitler for the duration of the war and then returning as quickly as possible to civilian life. John von Neumann (1903–1957) was a warrior who joined the game for life. The von Neumann era saw digital computers advance from breaking ciphers to guiding missiles and building bombs. The advent of the cold war was closely associated with the origins of high-speed electronic computers, whereby the power of new weapons could be proved by calculation instead of by lighting a fuse and getting out of the way.

Von Neumann had a gift for calculating the incalculable. After bringing thermonuclear Armageddon within reach, he applied his imagination to the possibility of certain especially cold-blooded forms of life. With his theory of self-reproducing automata he endowed Turing's Universal Machine with the power of constructing an unlimited number of copies of itself. Despite the threatening nature of these accomplishments, von Neumann was not an evil genius at heart. He was a mathematician who could not resist pushing the concepts of destruction and construction to their logical extremes, seeking to assign probabilities rather than moral judgment to the results.

Von Neumann played an enthusiastic role in the development of thermonuclear weapons, ballistic missiles, the application of game theory to nuclear deterrence, and other known and unknown black arts. He was one of the few Manhattan Project scientists who was not sequestered at Los Alamos, appearing periodically, like a comet, in the course of his transcontinental rounds. Advocating a hard line against the Soviet Union and publicly favoring a preventive nuclear attack, his views on nuclear war were encapsulated in his 1950 motto "Not whether but when." Nonetheless, he helped construct a policy of peace through the power of assured destruction that has avoided nuclear war for fifty years. Von Neumann's statements must be viewed not only in historical perspective, but also in the context of his pioneering work in game theory, which demonstrated the possibility of stabilizing a dangerously unstable situation by a convincing bluff—if and only if there appears to be the determination to back it up.

"Von Neumann seemed to admire generals and admirals and got along well with them," recalled Stan Ulam (1909–1984), a friend and colleague who shared in the full spectrum of von Neumann's work.[2] When von Neumann was invited to join the club and become a professional cold warrior, he did. He compressed an entire career as a military strategist into the final decade of his life. For the last nine months of von Neumann's battle with cancer, President Eisenhower arranged for a private suite at Walter Reed Hospital in Washington, D.C., assigning air force colonel Vincent Ford and eight airmen with top-secret clearance to provide twenty-four-hour protection and support.

When his end neared, von Neumann spoke not in the language of military secrets but in the Hungarian of his youth. Janos von Neumann was born in Budapest in 1903, the son of Max Neumann, a successful banker and economist elevated to the nobility by Emperor Franz Joseph in 1913. Life in the von Neumann household exposed young Johnny not only to economic theory but to administrative and political skills. Max brought his children into as much contact as possible with his world. "Managing a bank became a matter for family discussions, no less than our school subjects," recalled von Neumann's younger brother Nicholas. "All of us, but particularly John, observed and eventually used father's business techniques."[3] Nicholas remembers that his father, after making an investment in a textile works, brought home a card-controlled Jacquard automatic loom and believes that his brother's fascination with this mechanism resurfaced later in electronic form.

Von Neumann authored his first mathematical paper (with Michael Fekete, tutor turned collaborator) at the age of seventeen, launching a streak of productivity that continued without interruption until his death at age fifty-four—and for a period thereafter, if one includes his *Theory of Self-Reproducing Automata* (1966), reconstructed by logician Arthur Burks from von Neumann's unfinished manuscripts and notes. Von Neumann's *Mathematical Foundations of Quantum Mechanics* (1932) and *Theory of Games and Economic Behavior* (1944) remain classics in their fields. Eugene Wigner, a colleague since his school days in Budapest, commented that "nobody knows all science, not even von Neumann did. But as for mathematics, he contributed to every part of it except number theory and topology. That is, I think, something unique."[4] Von Neumann's mental acrobatics were legendary. "If you enjoy thinking, your brain develops," said Edward Teller. "And that is what von Neumann did. He enjoyed the functioning of his brain. And that is why he outdid anyone I know."[5] Hungary produced an exceptional crop of scientific talent between World War I and World War II. Von Neumann, Teller, Wigner, and Leo Szilard left their generation of mathematical physicists wondering how one small country had spawned four such minds at once. According to Ulam, von Neumann credited "the necessity of producing the unusual or facing extinction"[6]—a response that von Neumann pushed to its extreme. "Perhaps the consciousness of animals is more shadowy than ours and perhaps their perceptions are always dreamlike," wrote Eugene Wigner in 1964. "On the opposite side, whenever I talked with the sharpest intellect whom I have known—with von Neumann—I always had the impression that only he was fully awake, that I was halfway in a dream."[7]

Von Neumann saw his homeland disfigured by two world wars and a succession of upheavals in between. "I am violently anti-communist," he declared on his nomination to membership in the Atomic Energy Commission in 1955, "in particular since I had about a three-months taste of it in Hungary in 1919."[8] During the communist takeover the family retreated to the Italian Adriatic and was never personally at risk. Von Neumann spent the years 1921 to 1926 as a student shuttling between the University of Budapest, the University of Berlin, and the Eidgenössische Technische Hochschule (Federal Institute of Technology, or ETH) in Zurich, receiving both a degree in chemical engineering (assuring a livelihood) and a Ph.D. in mathematics (a field in which European positions were scarce). For the 1926–1927 academic year he received a Rockefeller Fellowship to work with David Hilbert at Göttingen, developing an axiomatization of set theory in support of Hilbert's program to formalize all mathematics from the ground up. Later, von Neumann admitted having had doubts that might have led him to anticipate Gödel's incompleteness results. Also with Hilbert, in 1927, he began his mathematical treatment of quantum mechanics, erecting a landmark in a field in which both geniuses and artisans were at work. He published twenty-five papers between 1926 and 1929. After accepting a visiting position at Princeton University in 1930, he was appointed to a full professorship there in 1931.

His escape from war-torn Europe left von Neumann determined to ensure that the most powerful weapons imaginable were placed in the hands of his adopted side. Along with fellow hydrogen-bomb designer Edward Teller, he perceived the Soviet threat to be only a stone's throw away—and the two Hungarians had seen what happened to the defenseless villages of their youth. "I don't think any weapon can be too large,"[9] he counseled Oppenheimer, who suffered second thoughts about atomic weapons, whereas von Neumann never flinched.

When von Neumann announced in 1946 that he considered bombs less important than computers, this did not mean that his interest in bombs had been eclipsed. He was thinking about both. The first job performed by the ENIAC, arranged under the auspices of von Neumann, was a feasibility study for the hydrogen, or super, bomb. To define the boundary conditions for the job, half a million IBM cards were shipped from Los Alamos to Philadelphia, where the calculation consumed six weeks and many more cards between November 1945 and January 1946. "This exposure to such a marvelous machine," recalled Nicholas Metropolis, who supervised the calculation, "coupled in short order to the Alamogordo [bomb test] experience was so singular that it was difficult to attribute any reality to either."[10] The shakedown run of the Institute for Advanced Study (IAS) computer, in the summer of 1951, was also a thermonuclear calculation for Los Alamos, running continuously for sixty days, well in advance of the machine's public dedication in 1952. "When the hydrogen bomb was developed," von Neumann testified at the Oppenheimer hearings in 1954, "heavy use of computers was made [but] they were not yet generally available . . . it was necessary to scrounge around and find a computer here and find a computer there which was running half the time and try to use it."[11] Ralph Slutz, who worked on the early stages of the IAS computer and then went on to construct the Bureau of Standards' SEAC, remembers "a couple people from Los Alamos" showing up as soon as the computer began operating, "with a program which they were terribly eager to run on the machine . . . starting at midnight, if we would let them have the time."[12]

Von Neumann believed that all fields of science, including pure mathematics, derive their sustenance through contact with real problems in the physical world. The military often arrives at these intersections first. Whether the application is for good or evil has little to do with the beauty of the science underneath. "If science is not one iota *more* divine for helping society, maybe she isn't one iota *less* divine for harming society," he wrote in 1954. "The principle of laissez faire has led to strange and wonderful results."[13]

In the United States, ballistics research fell under the domain of the U.S. Army's proving ground at Aberdeen, Maryland, founded in 1918, when field artillery was still being dragged around with horses but was beginning to fire high-velocity, long-range shells. Increasingly distant and mobile targets, especially aircraft, were difficult to hit by trial-and-error adjustment of the range. The converse presented an equivalent problem: how to hit a fixed target by dropping a bomb from a moving plane. Firing tables, tabulating target distance as a function of muzzle altitude, atmospheric conditions, temperature, and a host of other entangled variables became an essential adjunct to every gun. But their preparation required enormous numbers of complex calculations, largely performed by hand. The task resembled preparing the annual nautical almanac, except that it was necessary to prepare a separate almanac for each gun.

Mathematician Oswald Veblen (1880–1960), the proving ground's first director, assembled a notable constellation of mathematicians, including Norbert Wiener, at Aberdeen during World War I. The group dispersed its talent widely, contributing to every facet of computational mathematics and computer technology between World War I and World War II. Veblen became department head at Princeton University, soon making Princeton the rival of Göttingen in mathematics. In 1924, Veblen wrote a proposal for a Princeton mathematics institute that served, six years later, as a model for the creation of the Institute for Advanced Study, where both von Neumann and Veblen were appointed to professorships for life. During World War II, Veblen returned as chief scientist to the proving ground at Aberdeen, and, after von Neumann was naturalized as a U.S. citizen in 1937, Veblen recruited him to the Ballistic Research Laboratory's scientific advisory board. Advances in weaponry had left the principles of war unchanged, including the long-standing tradition of calling in the mathematicians to help aim catapults or guns.

"The perturbation arising from the resistance of the medium . . . does not, on account of its manifold forms, submit to fixed laws and exact description," Galileo explained in 1638, "and disturbs [the trajectory] in an infinite variety of ways corresponding to the infinite variety in the form, weight, and velocity of the projectiles."[14] Galileo found the behavior of high-velocity cannon fire mathematically impenetrable and limited his services to the preparation of firing tables for low-velocity trajectories, "since shots of this kind are fired from mortars using small charges and imparting no supernatural momentum they follow their prescribed paths very exactly."[15] The production of firing tables still demanded equal measures of art and science when von Neumann arrived on the scene at the beginning of World War II. Shells were test-fired down a range equipped with magnetic pickup coils to provide baseline data. Then the influence of as many variables as could be assigned predictable functions was combined to produce a firing table composed of between two thousand and four thousand individual trajectories, each trajectory requiring about 750 multiplications to determine the path of that particular shell for a representative fifty points in time.

A human computer working with a desk calculator took about twelve hours to calculate a single trajectory; the electromechanical differential analyzer at the Ballistic Research Laboratory (a ten-integrator version of the machine that Vannevar Bush had developed at MIT) took ten or twenty minutes. This still amounted to about 750 hours, or a month of uninterrupted operation, to complete one firing table. Even with double shifts and the assistance of a second, fourteen-integrator differential analyzer (constructed at the Moore School of Electrical Engineering at the University of Pennsylvania in Philadelphia), each firing table required about three months of work. The nearly two hundred human computers working at the Ballistic Research Laboratory were falling hopelessly behind. "The number of tables for which work has not been started because of lack of computational facilities far exceeds the number in progress," reported Herman Goldstine in August 1944. "Requests for the preparation of new tables are being currently received at the rate of six per day."[16]

Electromechanical assistance was not enough. In April 1943, the army initiated a crash program to build an electronic digital computer based on decimal counting circuits made from vacuum tubes. Rings of individual flip-flops (two vacuum tubes each) were formed into circular, ten-stage counters linked to each other and to an array of storage registers, forming, in effect, the electronic equivalent of Leibniz's stepped reckoner, but running at about six million rpm. The ENIAC (Electronic Numerical Integrator and Computer) was constructed by a team that included John W. Mauchly, John Presper Eckert, and (Captain) Herman H. Goldstine, supervised by John G. Brainerd under a contract between the army's Ballistic Research Laboratory and the Moore School. A direct descendant of the electromechanical differential analyzers of the 1930s, the ENIAC represented a brief but fertile intersection between the otherwise diverging destinies of analog and digital computing machines. Incorporating eighteen thousand vacuum tubes operating at 100,000 pulses per second, the ENIAC consumed 150 kilowatts of power and held twenty 10-digit numbers in high-speed storage. With the addition of a magnetic-core memory of 120 numbers in 1953, the working life of the ENIAC was extended until October 1955.

Programmed by hand-configured plugboards (like a telephone switchboard) and resistor-matrix function tables (set up as read-only memory, or ROM), the ENIAC was later adapted to a crude form of stored-program control. Input and output was via standard punched-card equipment requisitioned from IBM. It was thus possible for the Los Alamos mathematicians to show up with their own decks of cards and produce intelligible results. Rushed into existence with a single goal in mind, the ENIAC became operational in late 1945 and just missed seeing active service in the war. To celebrate its public dedication in February 1946, the ENIAC computed a shell trajectory in twenty seconds—ten seconds faster than the flight of the shell and a thousand times faster than the methods it replaced. But the ENIAC was born with time on its hands, because the backlog of firing-table calculations vanished when hostilities ceased.

Sudden leaps in biological or technological evolution occur when an existing structure or behavior is appropriated by a new function that spreads rapidly across the evolutionary landscape, taking advantage of a head start. Feathers must have had some other purpose before they were used to fly. U-boat commanders appropriated the Enigma machine first developed for use by banks. Charles Babbage envisioned using the existing network of church steeples that rose above the chaos of London as the foundation for a packet-switched communications net. According to neurophysiologist William Calvin, the human mind appropriated areas of our brains that first evolved as buffers for rehearsing and storing the precise timing sequences required for ballistic motor control. "Throwing rocks at even stationary prey requires great precision in the timing of rock release from an overarm throw, with the 'launch window' narrowing eight-fold when the throwing distance is doubled from a beginner's throw," he observed in 1983. "Paralleled timing neurons can overcome the usual neural noise limitations via the law of large numbers, suggesting that enhanced throwing skill could have produced strong selection pressure for any evolutionary trends that provided additional timing neurons. . . . This emergent property of parallel timing circuits has implications not only for brain size but for brain reorganization, as another way of increasing the numbers of timing neurons is to temporarily borrow them from elsewhere in the brain."[17] According to Calvin's theory, surplus off-hours capacity was appropriated by such abstractions as language, consciousness, and culture, invading the neighborhood much as artists colonize a warehouse district, which then becomes the gallery district as landlords raise the rent. The same thing happened to the ENIAC: a mechanism developed for ballistics was expropriated for something else.

John Mauchly, Presper Eckert, and others involved in the design and construction of the ENIAC had every intention of making wider use of their computer, but it was von Neumann who carried enough clout to preempt the scheduled ballistics calculations and proceed directly with a numerical simulation of the super bomb. The calculation took a brute-force numerical approach to a system of three partial differential equations otherwise resistant to analytical assault. As one million mass points were shuffled one IBM card at a time through the accumulators and registers at the core of the ENIAC's eighty-foot-long vault, the first step was taken toward the explosion of a device that was, as Oppenheimer put it, "singularly proof against any form of experimental approach."[18] The test gave a false positive result: the arithmetic was right, but the underlying physics was wrong. Teller and colleagues were led to believe that the super design would work, and the government was led to believe that if the Americans didn't build one the Soviets might build one first. By the time the errors were discovered, the hydrogen-bomb project had acquired the momentum to keep going until an alternative was invented that worked.

The super fizzled, but the ENIAC was hailed as an unqualified success. Because of the secret nature of their problem the Los Alamos mathematicians had to manage the calculation firsthand. They became intimately familiar with the operation of the computer and suggested improvements to its design. A machine capable of performing such a calculation could, in principle, compute an answer to any problem presented in numerical form. Von Neumann discovered in the ENIAC an instrument through which his virtuoso talents could play themselves out to the full—inventing new forms of mathematics as he went along. "It was his feeling that a mathematician who was pursuing some new field of endeavor or trying to extend the scope of older fields, should be able to obtain clues for his guidance by using an electronic digital machine," explained Willis Ware in 1953. "It was, therefore, the most natural thing that von Neumann felt that he would like to have at his own disposal such a machine."[19]

During the war, von Neumann had worked with the bomb designers at Los Alamos as well as with conventional-weapon designers calculating ballistic trajectories, blast and shock-wave effects, and the design of shaped charges for armor-piercing shells. It was his experience with the mathematics of shaped charges, in part, that led to the original success of the implosion-detonated atomic bomb. Bombs drew von Neumann's interest toward computers, and the growing power of computers helped sustain his interest in developing more powerful bombs. "You had an explosion a little above the ground and you wanted to know how the original wave would hit the ground, form a reflected wave, then combine near the ground with the original wave, and have an extra strong blast wave go out near the ground," recalled Martin Schwarzschild, the Princeton astrophysicist whose numerical simulations of stellar evolution had much in common—and shared computer time—with simulations of hydrogen bombs. "That was a problem involving highly non-linear hydrodynamics. At that time it was only just understood descriptively. And that became a problem that I think von Neumann became very much interested in. He wanted a real problem that you really needed computers for."[20]

Software—first called "coding" and later "programming"—was invented on the spot to suit the available (or unavailable) machines. Physicist Richard Feynman served on the front lines in developing the computational methods and troubleshooting routines used at Los Alamos in early 1944, when desk calculators and punched-card accounting machines constituted the only hardware at hand. The calculations were executed by dozens of human computers ("girls" in Feynman's terminology) who passed intermediate results back and forth, weaving together a long sequence, or algorithm, of simpler steps. "If we got enough of these machines in a room, we could take the cards and put them through a cycle. Everybody who does numerical calculations now knows exactly what I'm talking about, but this was kind of a new thing then—mass production with machines."

The problem was that the punched-card machines that Stan Frankel had ordered from IBM were delayed. So to test out Frankel's program, explained Feynman, "we set up this room with girls in it. Each one had a Marchant [mechanical calculator]. . . . We got speed with this system that was the predicted speed for the IBM machine[s]. The only difference is that the IBM machines didn't get tired and could work three shifts."[21] By keeping all stages of the computation cycle busy all the time, Feynman invented the pipelining that has maximized the performance of high-speed processors ever since.

Many thriving computer algorithms are direct descendants of procedures worked out by human computers passing results back and forth by hand. The initial challenge was how to break large problems into smaller, computable parts. Physically distinct phenomena often proved to be computationally alike. A common thread running through many problems of military interest was fluid dynamics—a subject that had drawn von Neumann's attention by its mathematical intractability and its wide-ranging manifestations in the physical world. From the unfolding of the afternoon's weather to the flight of a missile through the atmosphere or the explosion, by implosion, of atomic bombs, common principles are at work. But the behavior of fluids in motion, however optically transparent, remained mathematically opaque. Beginning in the 1930s, von Neumann grew increasingly interested in the phenomenon of turbulence. He puzzled over the nature of the Reynolds number, a nondimensional number that characterizes the transition from laminar to turbulent flow. "The internal motion of water assumes one or other of two broadly distinguishable forms," reported Osborne Reynolds in 1883, "either the elements of the fluid follow one another along lines of motion which lead in the most direct manner to their destination, or they eddy about in sinuous paths the most indirect possible."[22]

"It seemed, however, to be certain if the eddies were owing to one particular cause, that integration [of Stokes equations of fluid motion] would show the birth of eddies to depend on some definite value of $c\rho U/\mu$," explained Reynolds, introducing the parameter that bears his name.[23] As the product of length (of an object moving through a fluid or the distance over which a moving fluid is in contact with an object or a wall), density (of the fluid), and velocity (of the fluid or the object) divided by viscosity (of the fluid), the Reynolds number signifies the relative influence of these effects. All instances of fluid motion—water flowing through a pipe, a fish swimming through the sea, a missile flying through the air, or air flowing around the earth—can be compared on the basis of their Reynolds numbers to predict the general behavior of the flow. A low Reynolds number indicates the predominance of viscosity (owing to molecular forces between individual fluid particles) in defining the character of the flow; a high Reynolds number indicates that inertial forces (due to the mass and velocity of individual particles) prevail. Reynolds identified this pure, dimensionless number as the means of distinguishing between laminar (linear) and turbulent (nonlinear) flow and revealed how (but not why) the development of minute, unstable eddies precipitates self-sustaining turbulence as the transitional value is approached. The critical Reynolds number thus characterizes a transition between an orderly, deterministic regime and a disorderly, probabilistic regime that can be described statistically but not in full detail.

"Von Neumann ... wanted to find an explanation or at least a way to understand this very puzzling large number," wrote Ulam. "Small numbers like π and e, are of course very frequent in physics, but here is a number of the order of thousands and yet it is a pure number with no dimensions: it does whet our curiosity."[24] Von Neumann later suggested a similar distinction in computational complexity, marking the transition from a relatively small number of units forming an orderly, deterministic system to a probabilistic system composed of a large number of interconnected components whose behavior cannot be described (or predicted) more economically than by a statistical description of the system as a whole. Von Neumann was intrigued by the origins of self-organization in complicated systems—behavior reminiscent of the origins of turbulence but on a different scale. He understood that the boundaries between physics and computational models of physics are imprecise. The behavior of a turbulent hydrodynamic system can be predicted only by accounting for all interactions down to molecular scale. The situation can be modeled in computationally manageable form either by adopting a coarser numerical mesh, following a random sample of elements and drawing statistical conclusions accordingly, or by slowing the computation down in time. To make useful predictions of an ongoing process—say, a forecast of tomorrow's weather—the computation has to be speeded up.

The goal of weather prediction stimulated the development of electronic computers on several fronts. John Mauchly first conceived the ENIAC as an electronic differential analyzer to assist in recognizing long-term cycles in the weather—before another war came along with its demands for firing tables for bigger guns. Vladimir Zworykin (1889–1982), the brilliant Russian immigrant who brought electronic television into existence with his invention of the iconoscope, or camera tube, in 1923, also foresaw the potential of an electronic computer as a meteorological oracle for the world. Norbert Wiener, patron of cybernetics, who embraced the anti-aircraft gun but shunned the bomb, was a vocal proponent of atmospheric modeling and prophesied growing parallels between computers powerful enough to model nonlinear systems, such as the weather, and "the very complicated study of the nervous system which is itself a sort of cerebral meteorology."[25]

A cellular approach to numerically modeling the weather was developed by meteorologist Lewis Fry Richardson (1881–1953), who refined his atmospheric model, calculating entirely by hand and on "a heap of hay in a cold rest billet," while serving in the Friends' Ambulance Unit attached to the Sixteenth Division of the French infantry in World War I. "During the battle of Champagne in April 1917," wrote Richardson in the preface to his *Weather Prediction by Numerical Process* (1922), "the working copy was sent to the rear, where it became lost, to be rediscovered some months later under a heap of coal."[26] Richardson imagined partitioning the earth's surface into several thousand meteorological cells, relaying current observations to the arched galleries and sunken amphitheater of a great hall, where some 64,000 human computers would continuously evaluate the equations governing each cell's relations with its immediate neighbors, constructing a numerical model of the earth's weather in real time. "Perhaps some day in the dim future, it will be possible to advance the computations faster than the weather advances," hoped Richardson, "and at a cost less than the saving to mankind due to the information gained."[27]

Richardson thereby anticipated massively parallel computing, his 64,000 mathematicians reincarnated seventy years later as the multiple processors of Danny Hillis's Connection Machine. "We had decided to simplify things by starting out with only 64,000 processors," explained Hillis, recalling how Richard Feynman helped him bring Lewis Richardson's fantasy to life.[28] Even without the Connection Machine, Richardson's approach to cellular modeling would be widely adopted once it became possible to assign one high-speed digital computer rather than 64,000 human beings to keep track of numerical values and relations among individual cells. The faint ghost of Lewis Richardson haunts every spreadsheet in use today.

After the war, Richardson settled down as a meteorologist at Benson, Oxfordshire, contributing to the mathematical theory of turbulence and developing a novel method for remote sensing of movement in the upper air. By shooting small steel balls (between the size of a pea and a cherry) at the zenith and observing where they fell, Richardson helped turn swords into plowshares with a system that was faster, more accurate, and more robust than using balloons. When the Meteorological Office was transferred to the jurisdiction of the Air Ministry, Department of War, Richardson felt compelled, as a Quaker, to resign his post. Later still, when he discovered that poison-gas technicians were interested in his methods for predicting atmospheric flow, he ended his meteorological research, launching a mathematical investigation into the causes of war to which he devoted the remainder of his life. His studies were published posthumously in two separate volumes: *Arms and Insecurity,* an analysis of arms races, and *Statistics of Deadly Quarrels,* which documents every known category of violent conflict, from murder to strategic bombing, arranged both chronologically and on a scale of magnitude based on the logarithm of the number of victims left dead.[29]

Richardson, having managed an electrical laboratory before World War I, might have contributed more to the development of electronic computers had there been laboratory facilities not directly involved in military research. Working entirely on his own in the late 1920s in Paisley, Scotland, he produced an odd but insightful paper. "The Analogy Between Mental Images and Sparks" includes schematic diagrams of two simple electronic devices that Richardson constructed to illustrate his theories on the nature of synaptic function deep within the brain. One of these circuit diagrams is captioned "Electrical Model illustrating a Mind having a Will but capable of only Two Ideas."[30] Richardson had laid the foundations for massively parallel computing in the absence of any equipment except his own imagination; now, with nothing but a few bits of common electrical hardware, he gave bold hints as to the physical basis of mind. But he had no interest in elaborating on these principles or attempting to embody them on a wider scale. His ideas lay dormant in the pages of the *Psychological Review.*

Von Neumann saw to it that the powers of the electronic computer brought Richardson's dream (and, with the invention of the atomic bomb, his nightmares) to life. The first public announcement of von Neumann's postwar computer project was made by the *New York Times* after a meeting between Zworykin, von Neumann, and Francis W. Reichelderfer, chief of the U.S. Weather Bureau in Washington, D.C. The "development of a new electronic calculator, reported to have astounding potentialities . . . might even make it possible to 'do something about the weather,'" the *Times* reported. "Atomic energy might provide a means for diverting, by its explosive power, a hurricane before it could strike a populated place."[31] Stan Ulam hinted at the required scale: "To be used in 'weather control,' one will have to consider among other problems the interaction between several, perhaps nearly simultaneous, explosions."[32]

The ENIAC was still a military secret, leading the *Times* to conclude that "none of the existing [computing] machines, however, is as pretentious in scope as the von Neumann–Zworykin device." It *was* true that no program as ambitious was in the works. Von Neumann and Zworykin were not proposing to build just *a* computer, but a *network* of computers that would span the world. "With enough of these machines (100 was mentioned as an arbitrary figure) area stations could be set up which would make it possible to forecast the weather all over the world."[33]

Richardson's methods—breaking up a complex problem into a mosaic of computational cells—were equally adaptable to meteorology, fluid dynamics, and the peculiar shock-wave effects that governed both the construction of an atomic bomb and the physical destruction produced when one went off. Von Neumann took care of the bombs first. Later, when he developed his own computing center at the Institute for Advanced Study, he established a numerical meteorological group under Jule Charney that transformed Richardson's proposal into a working operation, leading directly to the system of numerical weather forecasting that models our atmosphere today.

The idea of building a general-purpose electronic digital computer had long been incubating in von Neumann's mind. "Von Neumann was well aware of the fundamental importance of Turing's paper of 1936 'On computable numbers' which describes in principle the 'Universal Computer' of which every modern computer (perhaps not ENIAC as first completed but certainly all later ones) is a realization," recalled Stan Frankel, who supervised numerical computation at Los Alamos during the war. "Von Neumann introduced me to that paper and at his urging I studied it with care. Many people have acclaimed von Neumann as the 'father of the computer' (in a modern sense of the term) but I am sure that he would never have made that mistake himself."[34] Separated by personality and style, Turing and von Neumann labored independently to bring digital computers to life. While Turing was at Princeton University in 1937, preparing his doctoral thesis under Alonzo Church, he worked in close proximity to von Neumann. But he declined the offer of a position as von Neumann's assistant, choosing instead to return to England and his destiny as the mastermind of Bletchley Park.

In contrast to the respectful distance that characterized his relations with Turing, von Neumann maintained a close friendship and long correspondence with Rudolf Ortvay, director of the Theoretical Physics Institute at the University of Budapest. There were two complementary approaches to digital computers. The first was to start from the most elementary beginnings, in the style of Leibniz or Turing, using nothing except 1s and 0s—or switches of some kind or another, which are 1s and 0s in physical form. The other approach, advocated by Ortvay, was to proceed in the opposite direction, taking as a starting point that most complicated known computer, the human brain.

"I read through your paper on games, and it gave me hope that you might succeed in formulating the problem of switching of brain cells if I succeed in drawing your attention to it," wrote Ortvay to von Neumann in 1941. Ortvay's suggestions encouraged von Neumann's efforts to develop a theory of automata general enough to apply both to the construction of digital computers and to the operation of the brain. "The brain can be conceived as a network with brain cells in its nodes. These are connected in a way that every individual cell can receive impulses from more than one other cell and can transmit impulses to several cells. Which of these impulses are received from or passed on to other cells may depend on the state of the cell, which in turn depends on the effects of anything that previously affected this particular cell. . . . The actual state of the cells (which I conceive as being numbered) would characterize the state of the brain. There would be a certain distribution corresponding to every spiritual state. . . . This model may resemble an automatic telephone switchboard; there is, however, a change in the connections after every communication."[35] The link between game theory and a theory of neural nets was never brought to fruition by von Neumann, although there are hints that his theory of automata was so inclined.

A Turing machine assembles a complex computation from a sequence of atomistic steps, whereas, as Ortvay suggested, the brain represents a computational process by a network of intercommunicating components, the chain of events being spatially distributed and not necessarily restricted to one computational step at a time. In 1943, neuropsychiatrist Warren S. McCulloch and mathematician Walter Pitts published their "Logical Calculus of the Ideas Immanent in Nervous Activity," showing that in principle (and for extremely simplified theoretical neurons) the computational behavior of any neural net can be duplicated exactly by an equivalent Turing machine.[36] This paper was widely cited in support of analogies between digital computers and brains. When von Neumann compiled the *First Draft of a Report on the EDVAC*, a document that launched the breed of stored-program computers that surround us today, he adopted the McCulloch–Pitts symbolism in diagramming the logical structure of the proposed computer and introduced terms such as *organ, neuron,* and *memory,* which were more common among biologists than among electrical engineers.

The EDVAC (Electronic Discrete Variable Automatic Computer) was conceived while the ENIAC was being built. The ability to modify its own instructions gained the EDVAC a distinction as the first full-fledged stored-program computer design. Beset by a series of technical and administrative delays, the original EDVAC did not become operational until late 1951, preceded and outperformed by its own more nimble offspring in both England and the United States. The EDVAC was nonetheless immortalized as the conceptual nucleus around which successive generations of computers formed. The project was initiated by Mauchly, Eckert, Goldstine, Arthur Burks, and others at the Moore School, but it was von Neumann's involvement that ignited the chain reaction that spread computers around the world. The EDVAC stored both data and instructions in mercury delay-line memory, as binary code. As in Turing's universal machine, long strings of bits defined not only numbers to be operated on but the sequence and potentially dynamic structure of the operations to be performed. The EDVAC thus embodied Turing's principle that complexity and adaptability could be more profitably assigned to the coding than to the machine.

Von Neumann consulted extensively with the EDVAC group in late 1944 and early 1945 and then, in a virtuoso performance, wrote up a detailed treatment of the engineering principles, logical architecture, and programming language ("order code") of the proposed computer, submitting the manuscript to the Moore School for review. Herman Goldstine had the incomplete draft (dated 30 June 1945) typed up, with von Neumann listed as sole author, and distributed it widely, with controversial results. On the one hand, the release of the EDVAC report parted the veil of secrecy that had obscured the ENIAC and Colossus projects during the war. The explicit instructions provided in the EDVAC report inspired a flurry of computer building and coding around the world—especially in England, where the Bletchley Park alumni remained handicapped by a prohibition against discussing their own existing work. On the other hand, the attribution of sole authorship to von Neumann embittered Eckert and Mauchly, who left the Moore School to found the Eckert–Mauchly Computer Company, producer of the BINAC and UNIVAC and ultimately acquired by Sperry-Rand. They believed, with some justification, that von Neumann's report had undermined their interest in future patents by placing the EDVAC design in the public domain. Insult was added to injury by von Neumann's eagerness to propagate the technology as widely and freely as possible, not only in conjunction with the government and academia, but also in cooperation with Eckert and Mauchly's competitors, such as RCA and IBM.

Von Neumann's computer project at the Institute for Advanced Study, launched at the end of 1945, received the bulk of its support not from industry but from the army, the navy, and the Atomic Energy Commission (AEC). Commercial benefits flowed mainly in reverse, with IBM and other organizations partaking freely of the IAS design and IAS-trained personnel. The Institute shied away from industrial contracts but had no qualms about accepting the support of Army Ordnance, the Office of Naval Research, and the AEC. Of some $772,000 of support for the IAS computer project between its inception in 1946 and June 1950, only $82,000 (excluding von Neumann's salary) was contributed by the IAS.[37]

Lewis Strauss, J. Robert Oppenheimer, and von Neumann all held influential positions at both the Institute and the AEC. Eventually this surfeit of influence presented a problem, because, as Herman Goldstine, administrative director of the computer project, explained, "when the Atomic Energy Commission up [and] decided one day that it was wrong for the Atomic Energy Commission to engage in research on electronic computers, we had nobody we could go to without all this fear of conflict of interest."[38] By the time AEC support of the IAS computer project wavered, the IAS design was being replicated widely and a derivative version (first known as the defense calculator) was being developed commercially by IBM. Originally targeted at defense contractors who were building new weapons systems, IBM renamed it the model 701 when they discovered the extent of the demand. Von Neumann was hired as a consultant, officially working thirty days a year for IBM.

Von Neumann's computer bore the paternity of war but, like the jet airplane or the Jeep, it did not remain exclusively a war machine for long. The defense industry employed the brightest minds of the time, individuals who commanded both a clear vision of the future of computers and the resources to bring this future about. "Over the years, the constant and most reliable support of computer science—and of science generally—has been the defense establishment," concluded Nicholas Metropolis and Gian-Carlo Rota in introducing a symposium on the history of digital computers at Los Alamos in 1976. "While old men in congresses and parliaments would debate the allocation of a few thousand dollars, farsighted generals and admirals would not hesitate to divert substantial sums to help oddballs in Princeton, Cambridge, and Los Alamos."[39] The oddballs turned out to be right.

The hydrogen bomb is a three-stage device. Thermonuclear fusion is triggered by a nuclear fission explosion, which is triggered by a high-explosive charge. With no room for trial and error, simulations executed by high-speed computers were as essential to successful bomb building as any of the other ingredients consumed—and transformed—along the way. Perhaps because of this close association from birth with bombs, electronic digital computers acquired an aura of explosiveness that lingers to this day. While computers were being used to catalyze this three-stage process, a series of repercussions was being reflected the other way. From one perspective, computers were testing bombs. From another perspective, bombs were testing computers, unleashing equally powerful results.

For fifty years, the bombs have remained under control. Our worst nightmare has become less of a nightmare as the century draws to a close. Of von Neumann's two creations, it is the computers that exploded, not the bombs.

6

RATS IN A CATHEDRAL

There is one further order that the control needs to execute. There should be some means by which the computer can signal to the operator when a computation has been concluded, or when the computation has reached a previously determined point. Hence an order is needed which will tell the computer to stop and to flash a light or ring a bell.

—BURKS, GOLDSTINE, AND VON NEUMANN[1]

I was eight years old, in 1961, when I stumbled on some relics of John von Neumann's electronic computer project left to molder away in an old barn. The barn was itself a relic, predating the establishment of the Institute for Advanced Study amid the Princeton, New Jersey, fields of Olden Farm. Littered with bales of hay, spring-toothed harrows, and other remnants of its working life, the barn now served as auxiliary storage to the Institute's physical plant and, on weekends, as a way station to a small band of boys who hunted frogs and turtles in the swamps and slow-moving streams that bordered the Institute woods. Ancient instincts drew us toward capturing small animals and dismantling machines. Our eyes adjusted to the darkness as a few rays of sunlight perforating the roof traced downward through the dust raised by the pigeons that fluttered away from us overhead. When we stopped talking, absolute silence reigned.

The old barn was a refuge to an extended family of ghosts. Something about abandoned machines—the suspension of life without immediate decay—evokes a mix of fear and hope. When the machine stops, we face whatever it is that separates death from life.

Our fathers were field theorists. At the Institute it was easier to find an expert in celestial mechanics than to find someone who worked on his or her own car. Agricultural implements were as foreign to us as were the mysterious contents of a series of heavy wooden crates piled in the center of the barn, filled with thick wood-and-metal plates of Mediterranean antiquities, the work of one

93

of the Institute's classical scholars awaiting a second printing that never came. After determining that the plates were the imprints of treasures and not treasures themselves, we scavenged on. Like so many grave robbers before us, we discovered that someone else had gotten to the good stuff first. At one end of the barn was a stockpile of war-surplus electronic equipment that had been selectively cannibalized for vacuum tubes and other vital parts. Partially eviscerated carcasses were distributed like livestock among the abandoned stalls.

We inspected cautiously; then we grew bold and returned with borrowed crescent wrenches and screwdrivers tucked under our belts. First we dismantled relays, making off with small electromagnets that we hooked up to battery power or doorbell transformers at home. Later we discovered microswitches: micro not in size but in the hair-trigger mechanism that shifted their internal state between off and on. Embedded within a maze of wiring and armored in Bakelite, they became the prized trophies of our hunt. Relays, solenoids, and microswitches were thoroughly intertwined. Relays were wired to solenoids wired to microswitches connected to other relays in turn, or sometimes back to the same relays once again. We blindly dissected the fossilized traces of electromechanical logic out of which the age of digital computers first took form. The primitive hardwired architecture, so accessible to our screwdrivers, remained impenetrable to our minds.

The square mile of fields and woodland surrounding the Institute for Advanced Study was cultivated, in lieu of forestry or agriculture, as a sanctuary for ideas. Its founders, in 1930, envisioned their educational utopia as a refuge from the mind-numbing bureaucracy of U.S. universities; they did not imagine the international upheaval from which their enclave would shortly offer an escape. "The Institute was a beacon in the descending darkness," wrote Director Harry Woolf in 1980, reflecting on the first fifty years, "a gateway to a new life, and for a very few a final place within which to continue to work and transmit to others the style and the techniques of great learning from the other shore."[2]

After the war the Institute became a permanent home to Albert Einstein, Kurt Gödel, John von Neumann, George Kennan, and other scholars equally distinguished if less well known. J. Robert Oppenheimer reigned as director from 1947 to 1966, presiding over what he described as an "intellectual hotel." He maintained the Institute's lead in mathematical physics while hosting transient scholars as diverse as child psychologist Jean Piaget and poet T. S. Eliot, a visiting member for the fall term of 1948 who listed *The Cocktail Party* (1950) as his only "publication related to IAS residence."[3] The Institute woods, bordered by the meandering bends of Stony Brook, offered sanctuary to indigenous wildlife as well, a refuge against the suburban fringe that was metastasizing up and down the eastern seaboard as inexorably as Dutch elm disease, consuming farmland as well as forest and leaving two-car garages in its wake.

The Institute for Advanced Study was conceived by Abraham Flexner (1866–1945), who rose from a Kentucky childhood as the son of an immigrant peddler to become first a schoolmaster and then an influential critic and leading reformer of higher education in the United States. Flexner credited his parents, Esther and Moritz Flexner, with being "shrewd enough to realize that their hold upon their children was strengthened by the fact that they held them with a loose rein."[4] This view formed the guiding principle of Flexner's educational career, even though "to be sure, we shall thus free some harmless cranks."[5] Academic freedom was not to be confused with a lowering of standards for academic work. Flexner emphasized "the impossibility of combining a tender regard for mediocrity with real enthusiasm for learning"[6] and secured his reputation as an educator, in 1887, by flunking an entire class.

The Institute appeared as a windfall late in Flexner's life. Flexner was sixty-four years old in 1930 and, as he recalled, "working quietly one day when the telephone rang and I was asked to see two gentlemen who wished to discuss with me the possible uses to which a considerable sum of money might be placed."[7] His visitors represented Louis Bamberger and his sister Caroline (Mrs. Felix) Fuld, retail merchants turned philanthropists after selling the Bamberger department store chain to R. H. Macy & Co., just in advance of the 1929 stock market crash. Flexner persuaded the Bambergers to underwrite not the medical college or local university they had originally intended but an institute that Flexner would later describe as "a paradise for scholars who, like poets and musicians, have won the right to do as they please."[8]

This was Flexner's chance to give substance to what he had been preaching about the deficiencies of higher education and research institutions for so many years. "Universities . . . are overorganized," was his main complaint. The Institute's goal was to avoid "dull and increasingly frequent meetings of committees, groups, or the faculty itself. Once started, this tendency toward organization and formal consultation could never be stopped."[9] The Institute for Advanced Study was incorporated on 20 May 1930, with Flexner as first director, followed by Frank Aydelotte in 1939 and J. Robert Oppenheimer in

1947. Albert Einstein and Oswald Veblen were appointed to the first professorships at the end of 1932, joined by John von Neumann, Hermann Weyl, and James Alexander in 1933. The School of Mathematics opened in 1933, followed by Humanistic Studies and Economics in 1935, Historical Studies in 1948, Natural Sciences in 1966, and Social Science in 1973. Theoretical Biology is rumored to be next.

"The Institute is, from the standpoint of organization, the simplest and least formal thing imaginable," explained Flexner. "Each school is made up of a permanent group of professors and an annually changing group of members. Each school manages its own affairs as it pleases; within each group each individual disposes of his time and energy as he pleases.... The results to the individual and to society are left to take care of themselves."[10] For its first decade the Institute had no buildings of its own and was incubated within Princeton University, establishing formal and informal relations between the two otherwise autonomous organizations that have continued to this day. "The mathematicians are guests of the Princeton mathematicians in Fine Hall," wrote Flexner in 1939 (from his own office among the dentists and lawyers of Nassau Street) and "the humanists are guests of the Princeton humanists in McCormick Hall," while, true to character (and overlooking the Springdale golf course), "the economists now occupy a suite at the Princeton Inn."[11]

Flexner believed in generous remuneration for the Institute's administration and permanent faculty, noting that although wealth might invite distractions from a scholar's academic work, "it does not follow that, because riches may harm him, comparative poverty aids him."[12] This generosity was not extended to visitors, "for on high stipends members will be reluctant to leave."[13] Flexner's stipulation that "no lines are drawn between professors, members, or visitors ... who mingle so freely as to be indistinguishable"[14] was upheld in the academic sphere, but in salary and housing the distinctions were made clear. Permanent faculty occupy substantial homes scattered among nearby streets and lanes, flanked by investment bankers, pharmaceutical heirs, successful New Jersey gangsters, and others who have managed to pay their dues. Rights of first refusal to these properties are retained by the Institute, extending its influence well beyond the eight hundred acres that make up its own grounds. Institute visitors, in contrast, are billeted in nondescript four-plex apartments, known locally as "the Project," circumscribing a communal launderette and some worn-out lawns. From a low hill (barely enough for winter sledding) Olden Manor, converted into the director's residence and surrounded by coach house, riding stables, and servants' quarters, overlooks the whole preserve.

The barn was situated below the project at the end of Olden Lane. It was the largest building on Olden Farm until eclipsed by the construction of Fuld Hall in 1939. An imposing, redbrick Georgian edifice, Fuld Hall housed the Institute's administrative offices, library, dining hall, several hallways leading to faculty offices in adjacent wings, and, on the ground floor, a common room presided over by a grandfather clock and leather-bound armchairs, with chessboards arranged near the windows that overlooked a courtyard facing the Institute woods. Fresh newspapers were skewered every morning on a polished wooden rack. Tea and cookies were served on real china daily at exactly three o'clock. Except for the absence of tennis courts (in the basement, table tennis was played underneath the steam ducts in a bare, concrete room), the resemblance to a private club or a European country estate (or perhaps a sanatorium) was pronounced. "We miss informal contact with one another and are about to remedy this defect by the erection of a building provided by the founders, to be called Fuld Hall,"[15] wrote Flexner in 1939. "Mathematicians will informally lunch, smoke, chat, walk, or play golf with the physicists and the director. Can any possible form of organization give the flexibility, the intimacy, the informality, the stimulus thus attainable?"[16] My father observed that although most university professors are parasites feeding on the living, Institute professors are saprophytes, feeding on the dead.

Between the barn and the housing project was a low brick building whose venetian blinds were always drawn. Graced with the appearance of a power substation or a telephone switching exchange, it was constructed suddenly, in 1947, to house the high-speed electronic computer whose circuits had begun to take shape in 1946 in the basement of Fuld Hall. Operational in 1951 and officially dedicated in 1952, the Institute for Advanced Study computer, or IAS machine, was never christened with an acronym of its own, but its far-flung offspring (ILLIAC and ORDVAC at the University of Illinois, JOHNNIAC at the Rand Corporation, MANIAC at Los Alamos National Laboratory, AVIDAC at Argonne, ORACLE at Oak Ridge, BESK in Stockholm, SILLIAC in Australia, BESM in Moscow, and WEIZAC in Israel, among some fifteen immediate siblings) were. After von Neumann's death in 1957 the project was wound down; the machine was turned over to Princeton University in 1958 and then dispersed, in part to the Smithsonian, in 1960. Although we Institute children competed in exploring every inch of the Institute's domain, from the abandoned pig farm at the far corner of the Institute woods to secret passageways in the basement of Fuld Hall, the computer building was out of bounds. In 1994 the building, still known as the

ECP building, was converted to a day-care center. Now a new generation is growing up in the same nursery where, by the flickering glow of twenty-six hundred vacuum tubes, the progenitor of the microprocessor was born.

After conflicts over patent rights divided the members of the ENIAC and EDVAC project at the Moore School, Eckert–Mauchly and von Neumann went separate ways. Von Neumann realized that the most direct way to influence the design of electronic computers, and to secure their unrestricted use, was to construct one for himself. "If he really wanted a computer," explained Arthur Burks, "the thing to do was to build it."[17] Von Neumann structured the project so as to introduce multiple copies of the new machine in several places at once. Progress reports were disseminated not only among the participating funding agencies and to a half-dozen groups that were duplicating the IAS design, but to any location where the potential of high-speed digital computers might fall on fertile ground. It is no accident that the vast majority of computers in circulation today follow the von Neumann architecture—characterized by a central processing unit operating in parallel on the multiple bits of one word of data at a time, a hierarchical memory ranging from fast but limited random-access memory to slow but unlimited media, such as floppy disks or tape, and a distinction between hardware and software that enabled robust computers (and a robust computer industry) to advance by a leapfrog process with each element evolving freely on its own.

At the time of von Neumann's project there were several competing architectures at the gate. Von Neumann chose, and spurred, the winning horse. "He was in the right place at the right time with the right connections with the right idea," explained Willis Ware, "setting aside the hassle that will probably never be resolved as to whose ideas they really were."[18] Although von Neumann did not foresee the personal computer, he did foresee, in agreement with Wiener's *Cybernetics*, that "science, as well as technology, will in the near and in the farther future increasingly turn from problems of intensity, substance, and energy, to problems of structure, organization, information, and control."[19]

MIT and the University of Chicago both offered substantial incentives to attract von Neumann, who remained determined to build his computer at the Institute for Advanced Study, despite the absence of laboratory facilities of any kind. "The Institute is a place which works with blackboards, paper and pencil, not with physical instruments and experimental techniques," recalled Willis Ware in 1953. "So the coming of six engineers with their assortment of oscilloscopes, soldering irons, and shop machinery was something of a shock."[20] In closing down the computer project after von Neumann's death, the trustees would reemphasize their commitment to pure research and advise that no experiments ever be permitted again. Institute colleagues, said Arthur Burks, "thought that von Neumann was not using his creative mathematical skills properly in being involved in computers."[21] Von Neumann was determined to prove them wrong. There were preliminary discussions with Princeton University concerning laboratory resources and staff support, but little materialized, although RCA's Princeton Laboratories made a firm commitment and did contribute technical support. With Aydelotte on his side, von Neumann was persuasive in his argument that the high-speed computer was a revolutionary instrument that would change the nature of mathematical research. In every field of science, von Neumann could cite specific instances in which a computational focus would give insight into the unknown. He had ready access to outside funding and was going to build a computer, either at the Institute or somewhere else. The trustees did not want to lose von Neumann to another institution, so they gave him the go-ahead.

In November 1945, a committee led by von Neumann and including Herman Goldstine (still attached to the ENIAC project at the Moore School) held its first meeting in Vladimir Zworykin's office at RCA. Also in attendance was John Tukey, professor of mathematics at Princeton (and the originator of information theorist Claude Shannon's landmark contraction of "binary digit" to "bit"). Von Neumann issued a memorandum of objectives, concluding that "it is to be expected that the future evolution of high-speed computing will be decisively influenced by the experiences gained."[22] By the spring of 1946 the project was under way and staff, led by Goldstine, were signing on.

Goldstine, Arthur Burks, and von Neumann set to work developing the logical plan of the computer, released in June 1946 as *Preliminary Discussion of the Logical Design of an Electronic Computing Instrument*, revised in September 1947, and followed by a three-volume report, *Planning and Coding of Problems for an Electronic Computing Instrument* (1947–1948). These documents were circulated even more widely than the EDVAC report, setting precedents that computer architecture and programming has followed ever since. "The remarkable feature of the reports," according to I. J. Good, "was that they gave lucid reasons for every design decision, a feature seldom repeated in later works."[23]

Von Neumann worked out of his office at the Institute, and Goldstine and Burks were installed in an annex to Gödel's office down the hall. "Kurt Gödel didn't have a secretary, didn't want one, I assume," recalled Burks. "So for that summer, when of course we didn't yet have a building for the computer, Herman and I occupied the secretary's office next to Gödel's office. It had a blackboard on the wall."[24] This was more than could be said for the facilities provided the engineers, who were given a bare storage room in the basement of Fuld Hall. "Our first job was to build work tables for us to work on," recalled Ralph Slutz, a Princeton graduate student who joined the group in June after completing his degree. "We asked von Neumann if he would pay for the paint if we painted the walls a more reasonable color than they were when we moved in. This he did."[25] Willis Ware, a twenty-eight-year-old electrical engineer who also arrived in June, recalled the "temporary space in the second basement, surrounding the boilers. You know—it wasn't bad since it was summer and they were turned off. And then we moved upstairs to the first basement under one wing."[26]

Von Neumann had no experience—or desire to experience—building anything with his own hands. Whether at a bomb test in the desert or at a meeting of the trustees, he invariably appeared in a three-piece suit, wearing the uniform of a banker among colleagues who formed a generally disheveled troop. "Once he understood the principle of it, the ghastly details like the fact that you'd have to put by-pass condensers on things, and all sorts of dirty engineering things—that didn't really interest him," noted Goldstine. "He would have made a lousy engineer."[27] To translate his logical plan into reality he needed the help of those who could design and build not only the logical control, arithmetic unit, and high-speed memory, but everything else from stable power supplies to air-conditioning to some means of reading data in and out of the machine. As chief engineer von Neumann selected Julian Bigelow, thirty-three years old and "a quiet, thorough New Englander" in the opinion of Norbert Wiener, with whom Bigelow had collaborated during the war on real-time computing for anti-aircraft fire control. Wiener recommended Bigelow for the job. "We telephoned from Princeton to New York, and Bigelow agreed to come down in his car. We waited till the appointed hour and no Bigelow was there. He hadn't come an hour later. Just as we were about to give up hope, we heard the puffing of a very decrepit vehicle. It was on the last possible explosion of a cylinder that he finally turned up with a car that would have died months ago in the hands of anything but so competent an engineer."[28]

Bigelow was a theoretician as well as a mechanic, and a founding member of the cybernetics group. With Norbert Wiener and Arturo

Rosenblueth he coauthored a 1943 paper, "Behavior, Purpose and Teleology," suggesting unifying principles underlying intelligent behavior among living beings and machines. "A further comparison of living organisms and machines ... may depend on whether or not there are one or more qualitatively distinct, unique characteristics present in one group and absent in the other," concluded Bigelow. "Such qualitative differences have not appeared so far."[29] This paper served as the namesake for the informal Teleological Society out of which the Macy conferences and what came to be known as the cybernetics movement took form. "Cybernetics came into its own," explained Warren McCulloch in 1961, "when Julian Bigelow pointed out the fact that it was only information concerning the outcome of the previous act that had to return."[30]

Bigelow, who published little, exercised his influence by linking the mathematical worlds of Wiener and von Neumann with the world of practical machines. He had a gift for building things that worked and fixing those that broke. He was perhaps the only permanent member of the Institute who worked on his own car. "I remember one day walking out the back door of that little brick building," remarked Ware, "and here's Julian lying under this little Austin, welding a hole in a gas tank. And he said, 'Nope! It won't explode!' And he had some perfectly reasonable explanation for why it wouldn't explode, based on the principles of physics."[31]

Bigelow's job was to take the logical design as laid out in the abstract by von Neumann and coax it to life as a machine. "Julian would have the ideas, or Ralph [Slutz] would kind of detail the ideas, and then Pom [James Pomerene] and I would go try and make the electrons do their thing," said Ware. "He was kind of more physicist and theoretician than engineer. ... In modern parlance, what you'd say was: Julian was the architect of that machine."[32] His responsibilities included designing and fabricating circuits, building test equipment that could track events in the microsecond range, requisitioning scarce electronic parts, and securing something even scarcer in Princeton—living quarters for his staff. Bigelow managed to find some war-surplus housing and had it moved on site, against the protests of Princeton citizens who felt this might bring the wrong kind of architecture to their side of the tracks. Using their small machine shop, Bigelow's crew constructed not only the main frame of the computer but peripheral equipment as well, such as a forty-four-track high-speed magnetic drum (storing 2,048 40-bit words) and a high-speed wire drive that coiled and uncoiled magnetic recording wire at up to one hundred feet (or 90,000 bits) per second from a pair of bicycle wheels, differentially coupled side by side on a single concentric drive. The wheels could be removed and inserted as a unit, just as a disk cartridge or other removable medium is used today.

Bigelow was joined by a young, talented, and enthusiastic crew of technicians and engineers, mingled with visiting mathematicians and scientists who were coding their own particular problems for the computer that had not yet been built. "A long chain of improbable chance events led to our involvement," recalled Bigelow. "People ordinarily of modest aspirations, we all worked so hard and selflessly because we believed—we knew—it was happening here and at a few other places right then, and we were lucky to be in on it. We were sure because von Neumann cleared the cobwebs from our minds as nobody else could have done. A tidal wave of computational power was about to break and inundate everything in science and much elsewhere, and things would never be the same afterward. It would cleanse and solve areas of obscurity and debate that had piled up for decades. Those who really understood what they were trying to do would be able to express their ideas as coded instructions, calculate with powerful machines, and find answers and demonstrate explicitly by numerical experiments. The process would advance and solidify knowledge and tend to keep men honest."[33]

The group worked hard and played hard, and despite (or because of) delays attributed to Bigelow's meticulous attention to detail the machine got finished and it worked. "The rate at which Julian could think, and the rate at which Julian could put ideas together was the rate at which the project went," Ware observed.[34] Although received coolly by the Institute ("We were doing things with our hands and building dirty old equipment. That wasn't the Institute," said Ware), the engineers were welcomed at von Neumann's home and treated to the hospitality that was a trademark of von Neumann's scientific career. My father, a visiting member at the Institute in 1948, remembers how the Institute's abstract and theoretical atmosphere was enlivened by "von Neumann and his band of freaks." Although Institute members were known for eccentric driving habits, Ware singled out one occasion when James Pomerene and Nick Metropolis drove home from one of the von Neumann gatherings in reverse.

In late 1946, the AEC agreed to provide funds for a plain concrete (and officially temporary) building to house the computer along with some modest facilities for its construction and operational support. The Institute agreed to provide a brick veneer, cautiously accepting the structure as an outlying satellite of Fuld Hall. Arthur Burks reminisced about "going with Herman [Goldstine] and Oswald

Veblen to pick a site for the new building. And we walked through the woods, but it was clear that Veblen didn't want any trees to be cut down. . . . In the end, he picked a site which was low down, not too far away from the Institute building so it wasn't inconveniently far away. He wanted the building to be one story only, so that this would not be a conspicuous building."[35]

The computer, on the other hand, *was* conspicuous—for the unprecedented power and economy of its design and the resourcefulness with which these principles were engineered. Von Neumann's mathematical vision was translated into the visible elegance of a machine. In physical appearance it resembled a turbocharged V-40 engine, about six feet high, two feet wide, and eight feet long. The computer weighed one thousand pounds; the air-conditioning unit weighed fifteen tons. Overhead ducts removed 52,000 Btu of waste heat per hour via a network of cooling channels that infiltrated the core of the machine. Some twenty-six hundred vacuum tubes were neatly arranged in a series of shift registers and accumulators that shuffled electrons through gates, toggles, and switches at up to one million cycles per second, executing precisely those binary processes that Leibniz had envisioned performing with marbles in 1679. The geometry was compact ("perhaps too compact for convenient maintenance," admitted Bigelow), but a minimal connection path between components was achieved by these convolutions in the chassis, like the folding of a cerebral cortex into a skull. The forty cylinders arranged in a bank of twenty along each side of the base of the machine contributed the driving force (and chief obstacle) to its design. They contained the world's first fully-random-access memory, or RAM. There were only 1,024 bits per cylinder, but with twenty-four-microsecond access time this was more horsepower than the young science of electronic digital processing had ever seen.

Digital computers, since the time of Babbage, had relied on serial memory (although the need for random access was recognized in the way in which Babbage's mechanical "store" of variables was to be made available to his arithmetic "mill"). No matter what the medium—paper tape, punched cards, or magnetic media—the processor shuffled through the contents of its memory in sequence, with associated delays. IBM's Selective Sequence Electronic Calculator (SSEC), completed in 1948 and housed in a windowed showroom on Fifty-seventh Street in New York, represented the punched-tape dinosaur against which the IAS machine would play the mouse that roared. The SSEC stored some twenty thousand 20-digit numbers on eighty-track paper tape, written by three punching units and referred to by a formidable array of sixty-six reading heads. Despite this ability to consult its memory in sixty-six places at once, access to a given location could take up to a second, impressive to onlookers but not to the future of IBM. Acoustic delay-line memory, though a thousand times faster, required ingenious coding and precise synchronization—a challenge similar to trying to play a game of cards while shuffling the deck.

In association with Vladimir Zworykin and Jan Rajchman at RCA, von Neumann arranged to develop a digital memory tube for the IAS computer, christened the Selectron. Information was written by an electron beam projected through an electromagnetic mask controlled by digital switching and read from an array of 4,096 separate targets (tiny, nickel-plated eyelets arranged like Cheerios on a mica sheet) that shifted state individually to store one bit of data (accessible at random) each. After two years, there were no Selectrons in existence ("They were doing things inside that vacuum that hadn't been done before," said Ware), although a 256-bit version was eventually produced in limited quantities and used successfully in the IAS-derived JOHNNIAC built at Rand. The IAS team decided to pursue its own alternative, using commercially available parts.

The IAS memory was based on the Williams tube—an ordinary cathode-ray tube (CRT) modified to allow data to be read, written, and continuously refreshed as a pattern of charged spots on the phosphor coating inside the tube. The state of an individual spot was distinguished by "interrogating" the spot with a brief pulse of electrons and noting the character of a very faint secondary current induced in a wire screen attached to the tube's outside face. Von Neumann had discussed the underlying concept—in principle similar to Zworykin's iconoscope but operating in reverse—while at the Moore School in 1944 and explored its possible use as a high-speed storage medium in the EDVAC report of 1945. Frederick C. Williams, after working on pulse-coded IFF (Identification Friend or Foe) radar systems at England's Telecommunications Research Establishment during the war, developed a practical version in 1946 and succeeded in building a small computer at Manchester University, under the direction of M. H. A. Newman, that demonstrated CRT-based storage and a rudimentary stored program in June 1948. The prototype operated in serial mode, cycling through the pattern of spots in a series of traces, like an oscilloscope or a television, thereby reading and writing the entire sequence of bits thousands of times per second—a vastly accelerated version of one of the loops of paper tape used by the Colossus at Bletchley Park. You could watch the bits of information dancing on the screen as a computation proceeded, and Turing, who soon joined the Manchester group, was noted for his ability to read numbers directly off the screen, just as he had been able to read binary code directly from teletypewriter tapes as intercepted messages were being sorted out.

It was evident (as had been recognized by Zworykin in the 1930s) that random access was possible if suitable control circuits for the electron-beam deflection voltages were engineered. Bigelow paid a visit to Manchester in June 1948, and the IAS team soon developed switching circuits that could read or write to any location at any time, appropriating a few microseconds before resuming the normal scanning and refresh cycles where they left off. The resulting memory organ was in effect an electronically switched 32 × 32 array of capacitors but was, as Bigelow noted, "one of mankind's most sensitive detectors of electromagnetic environmental disturbances."[36] The internal coating had to be flawless, and shielding had to be religiously maintained. RCA and one other manufacturer allowed the IAS to scan their inventory for unblemished specimens and ship the other 80 percent of them back. A forty-first monitor stage could be switched over to any of the forty memory stages, allowing the operator to inspect the contents of the memory to see how a computation was progressing—or why it had unexpectedly stopped. This was later augmented by a separate seven-inch cathode-ray tube serving as a 7,000-points-per-second graphical display.

All forty memory tubes had to work perfectly at the same time. Data were processed in parallel (not parallel processing as the term is used today) by operating on all the digits of a 40-bit word at once. The 40 bits represented either a number or a pair of 20-bit instructions, of which 10 bits designated the order and 10 bits a memory address. Each of the 40 bits making up a word was assigned the same position in a different Williams tube, an addressing scheme analogous to handing out similar room numbers in a forty-floor hotel. The forty Williams tubes were controlled in unison, like a bank of TV sets tuned to the same channel for display. This made the computer forty times as fast as a serial processor, but, in the opinion of numerous skeptics, unlikely to work without one small thing or another always going wrong. "The rig can be viewed as a big tube test rack," observed Bigelow, and it is remarkable that between the forty Williams tubes and twenty-six hundred other vacuum tube envelopes, the machine eventually worked more than 75 percent of the time.[37]

When Pomerene achieved a thirty-four-hour error-free test of a two-stage memory on July 28–29, 1949, the team knew their greatest obstacle had been solved. The rest of the computer could be built from standard components whose behavior was, for the most part, known. The arithmetic unit was kept as simple as possible: an accumulator, two shift registers, an adder, and a digit resolver. The core of the computer was essentially a very fast (thirty-one microsecond) adding machine. As Thomas Hobbes had pointed out in 1651, from simple addition (and the addition of a binary complement, which equals subtraction) one can, by careful bookkeeping, construct everything else. All the bits, represented by delicately balanced pulses of electrons, were forced to march cautiously, one step at a time. "Information was first locked in the sending toggle; then gating made it common to both sender and receiver, and then when securely in both, the sender could be cleared," Bigelow explained. "Information was never 'volatile' in transit; it was as secure as an acrophobic inchworm on the crest of a sequoia."[38] There was no floating-point arithmetic. The prospect was considered but rejected as not essential at the time. The programmer had to guess where the most significant digit ended up and test accordingly to "bring it back into focus" as the computation moved along. There were twenty basic instructions, with forty-four order codes. "During the spring of 1951, the machine became increasingly available for use, and programmers were putting their programs on for exploratory runs," said Bigelow. "The machine error rate had become low enough so that most of the errors found were in their own work."[39]

The original input and output to the computer was via five-hole paper teletypewriter tape, fed through a customized interface dubbed the "inscriber" and the "outscriber." It took almost thirty minutes to load 1,024 words, one register at a time, into the memory of the machine. After a few months of operation, a standard IBM 516 reproducing punch (designed to read and write 12-bit columns) was rewired to read 40 bits in parallel (every other punch position in an eighty-column row), allowing the memory to be filled in five minutes or less. Output could be punched at one hundred cards per minute, allowing a skilled operator "to interpret the perforations visually and so diagnose what was happening to his computation while away from the machine."[40] IBM's policy at the time allowed no customer modifications to its equipment. The exception granted to the Institute had consequences that were hardly envisioned at the time. The jury-rigged hybrid demonstrated at the Institute led directly to commercial production of the IBM 701, helping to secure leadership of the electronic data-processing industry for IBM.

Von Neumann circulated at the highest levels of the scientific and political establishment. Largely through his influence, the project was duplicated rapidly around the world. In the race to build working computers, the "few more months" that always remained until a particular machine would be up and running became known as the "von Neumann constant." It was the challenge of beating this famous constant, and the advantage of following rather than breaking the engineering trail, that led several groups—at the University of Illinois, the Bureau of Standards, Argonne National Laboratory, and Los Alamos—to get their machines running ahead of the official dedication ceremony (10 June 1952) at the IAS. "Many of us who are in the course of making copies of the IAS machine have a tendency to emphasize our deviations and forget the tremendous debt that we owe Julian Bigelow and others at the Institute," admitted William F. Gunning of RAND in 1953. "The fact that so many of us have been able to make an arithmetic unit that works when first plugged in . . . is proof enough."[41]

A constant stream of brilliant individuals—from distinguished scientists to otherwise unknowns—appeared in Princeton to run their problems on the IAS machine. For this the Institute was ideal. The administration was flexible, intimate, and spontaneous. The computer project operated on a shoestring compared to other laboratories but was never short of funds. The facilities were designed to accommodate visitors for a day, a month, or a year, and the resources of Princeton University were close at hand. There were no indigenous computer scientists monopolizing time on the machine, although a permanent IAS meteorological group under Jule Charney ran their simulations regularly and precedence was still granted to the occasional calculation for a bomb. "My experience is that outsiders are more likely to use the machine on important problems than is the intimate, closed circle of friends," recalled Richard Hamming, looking back on the early years of computing in the United States.[42]

The machine was duplicated, but von Neumann remained unique. His insights permeated everything that ran on the computer, from the coding of Navier–Stokes equations for compressible fluids to S. Y. Wong's simulation of traffic flow (and traffic jams) to the compilation of a historical ephemeris of astronomical positions covering the six hundred years leading up to the birth of Christ. "Quite often the likelihood of getting actual numerical results was very much larger if he was not in the computer room, because everybody got so nervous when he was there," reported Martin Schwarzschild. "But when you were in real thinking trouble, you would go to von Neumann and nobody else."[43]

Von Neumann's reputation, after fifty years, has been injured less by his critics than by his own success. The astounding proliferation of the von Neumann architecture has obscured von Neumann's contributions to massively parallel computing, distributed information processing, evolutionary computation, and neural nets. Because his deathbed notes for his canceled Silliman lectures at Yale were published posthumously (and for a popular audience) as *The Computer and the Brain* (1958), von Neumann's work has been associated with the claims of those who were exaggerating the analogies between the digital computer and the brain. Von Neumann, on the contrary, was preoccupied with explaining the differences. How could a mechanism composed of some ten billion unreliable components function reliably while computers with ten thousand components regularly failed?

Von Neumann believed that entirely different logical foundations would be required to arrive at an understanding of even the simplest nervous system, let alone the human brain. His *Probabilistic Logics and the Synthesis of Reliable Organisms from Unreliable Components* (1956) explored the possibilities of parallel architecture and fault-tolerant neural nets. This approach would soon be superseded by a development that neither nature nor von Neumann had counted on: the integrated circuit, composed of logically intricate yet structurally monolithic microscopic parts. Serial architecture swept the stage. Probabilistic logics, along with vacuum tubes and acoustic delay-line memory, would scarcely be heard from again. If the development of solid-state electronics had been delayed a decade or two we might have advanced sooner rather than later into neural networks, parallel architectures, asynchronous processing, and other mechanisms by which nature, with sloppy hardware, achieves reliable results.

Von Neumann was as reticent as Turing was outspoken on the question of whether machines could think. Edmund C. Berkeley, in his otherwise factual and informative 1949 survey, *Giant Brains*, captured the mood of the time with his declaration that "a machine can handle information; it can calculate, conclude, and choose; it can perform reasonable operations with information. A machine, therefore, can think."[44] Von Neumann never subscribed to this mistake. He saw digital computers as mathematical tools. That they were members of a more general class of automata that included nervous systems and brains did not imply that they could think. He rarely discussed artificial intelligence. Having built one computer, he became less interested in the question of whether such machines could learn to think and more interested in the question of whether such machines could learn to reproduce.

" 'Complication' on its lower levels is probably degenerative, that is, that every automaton that can produce other automata will only be able to produce less complicated ones," he noted in 1948. "There is, however, a certain minimal level where this degenerative characteristic ceases to be universal. At this point automata which can reproduce themselves, or even construct higher entities, become possible."[45] Millions of very large scale integrated circuits, following in the footsteps of the IAS design but traced in silicon at micron scale, are now replicated daily from computer-generated patterns by computer-operated tools. The newborn circuits, hidden in clean rooms and twenty-four-hour-a-day "fabs," where the few humans present wear protective suits for the protection of the machines, are the offspring of von Neumann's *Theory of Self-Reproducing Automata.* Just as predicted, these machines are growing more complicated from one generation to the next. None of these devices, although executing increasingly intelligent code, will ever become a brain. But collectively they might.

Von Neumann's Silliman lecture notes gave "merely the barest sketches of what he planned to think about," noted Stan Ulam in 1976. "He died so prematurely, seeing the promised land but hardly entering it."[46] Von Neumann may have envisaged a more direct path toward artificial intelligence than the restrictions of the historic von Neumann architecture suggest. High-speed electronic switching allows computers to explore alternatives thousands or even millions of times faster than biological neurons, but this power pales in comparison with the combinatorial abilities of the billions of neurons and uncounted synapses that constitute a brain. Von Neumann knew that a structure vastly more complicated, flexible, and *unpredictable* than a computer was required before any electrons might leap the wide and fuzzy distinction between arithmetic and mind. Fifty years later, digital computers remain rats running two-dimensional mazes at basement level below the foundations of mind.

As a practicing mathematician and an armchair engineer, von Neumann knew that something as complicated as a brain could never be designed; it would have to be evolved. To build an artificial brain, you have to grow a matrix of artificial neurons first. In 1948, at the Hixon Symposium on Cerebral Mechanisms in Behavior, von Neumann pointed out in response to Warren S. McCulloch that "parts of the organism can act antagonistically to each other, and in evolution it sometimes has more the character of a hostile invasion than of evolution proper. I believe that these things have something to do with each other." He then described how a primary machine could be used to exploit certain tendencies toward self-organization among a large number of intercommunicating secondary machines. He believed that selective evolution (via mechanisms similar to economic competition) of incomprehensibly complex processes among the secondary machines could lead to the appearance of comprehensible behavior at the level of the primary machine.

"If you come to such a principle of construction," continued von Neumann, "all that you need to plan and understand in detail is the primary automaton, and what you must furnish to it is a rather vaguely defined matrix of units; for instance, 10^{10} neurons which swim around in the cortex. . . . If you do not separate them . . . then, I think that it is achievable that the thing can be watched by the primary automaton and be continuously reorganized when the need arises. I think that if the primary automaton functions in parallel, if it has various parts which may have to act simultaneously and independently on separate features, you may even get symptoms of conflict . . . and, if you concentrate on marginal effects, you may observe the ensuing ambiguities. . . . Especially when you go to much higher levels of complexity, it is not unreasonable to expect symptoms of this kind."[47] The "symptoms of this kind" with which von Neumann and his audience of neurologists were concerned were the higher-order "ensuing ambiguities" that somehow bind the ingredients of logic and arithmetic into the cathedral perceived as mind.

Von Neumann observed, in 1948, that information theory and thermodynamics exhibited parallels that would grow deeper as the two subjects were mathematically explored. In the last years of his foreshortened life, von Neumann began to theorize about the behavior of populations of communicating automata, a region in which the parallels with thermodynamics—and hydrodynamics—begin to flow both ways. "Many problems which do not *prima facie* appear to be hydrodynamical necessitate the solution of hydrodynamical questions or lead to calculations of the hydrodynamical type," von Neumann had written in 1945. "It is only natural that this should be so."[48]

Lewis Richardson's sphere of 64,000 mathematicians would not only model the large-scale turbulence of the atmosphere, they might, if they calculated and communicated fast enough, acquire an atmosphere of turbulence of their own. As self-sustaining vortices arise spontaneously in a moving fluid when velocity outweighs viscosity by a ratio to which Osborne Reynolds gave his name, so self-sustaining currents may arise in a computational medium when the flow of information among its individual components exceeds the computational viscosity by a ratio that John von Neumann, unfortunately, did not live long enough to define.

7

SYMBIOGENESIS

Instead of sending a picture of a cat, there is one area in which we can send the cat itself.

—MARVIN MINSKY[1]

"During the summer of 1951," according to Julian Bigelow, "a team of scientists from Los Alamos came and put a large thermonuclear calculation on the IAS machine; it ran for 24 hours without interruption for a period of about 60 days, many of the intermediate results being checked by duplicate runs, and throughout this period only about half a dozen errors were disclosed. The engineering group split up into teams and was in full-time attendance and ran diagnostic and test routines a few times per day, but had little else to do. So it had come alive."[2]

The age of digital computers dawned over the New Jersey countryside while a series of thermonuclear explosions, led by the MIKE test at Eniwetok Atoll on 1 November 1952, corroborated the numerical results. In 1953, a series of experiments performed at the Institute for Advanced Study demonstrated that digital computers could be used not only to develop the means of destroying life, but to spawn lifelike processes of a form so far entirely unknown.

Italian-Norwegian mathematician Nils Aall Barricelli (1912–1993) arrived at the Institute as a visiting member for the spring term of 1953. Barricelli initiated extensive tests of evolution theories in 1953, 1954, and 1956, using the IAS computer to develop a working model of Darwinian evolution and to investigate the role of symbiogenesis in the origin of life.

The theory of symbiogenesis was introduced in 1909 by Russian botanist Konstantin S. Merezhkovsky (1855–1921) and expanded by Boris M. Kozo-Polyansky (1890–1957) in 1924.[3] "So many new facts arose from cytology, biochemistry, and physiology, especially of lower organisms," wrote Merezhkovsky in 1909, "that [in] an attempt once

111

again to raise the curtain on the mysterious origin of organisms . . . I have decided to undertake . . . a new theory on the origin of organisms, which, in view of the fact that the phenomenon of symbiosis plays a leading role in it, I propose to name the theory of symbiogenesis."[4] Symbiogenesis offered a controversial adjunct to Darwinism, ascribing the complexity of living organisms to a succession of symbiotic associations between simpler living forms. Lichens, a symbiosis between algae and fungi, sustained life in the otherwise barren Russian north; it was only natural that Russian botanists and cytologists took the lead in symbiosis research. Taking root in Russian scientific literature, Merezhkovsky's ideas were elsewhere either ignored or declared unsound, most prominently by Edmund B. Wilson's dismissal of symbiogenesis as "an entertaining fantasy . . . that the dualism of the cell in respect to nuclear and cytoplasmic substance resulted from the symbiotic association of two types of primordial microorganisms, that were originally distinct."[5]

Merezhkovsky viewed both plant and animal life as the result of a combination of two plasms: *mycoplasm*, represented by bacteria, fungi, blue-green algae, and cellular organelles; and *amoeboplasm*, represented by certain "monera without nuclea" that formed the nonnucleated material at the basis of what we now term eukaryotic cells. Merezhkovsky believed that *mycoids* came first. When they were eaten by later-developing *amoeboids* they learned to become nuclei rather than lunch. It is equally plausible that amoeboids came first, with mycoids developing as parasites later incorporated symbiotically into their hosts. The theory of two plasms undoubtedly contains a germ of truth, whether the details are correct or not. Merezhkovsky's two plasms of biology were mirrored in the IAS experiments by embryonic traces of the two plasms of computer technology—hardware and software—that were just beginning to coalesce.

The theory of symbiogenesis assumes that the most probable explanation for improbably complex structures (living or otherwise) lies in the association of less complicated parts. Sentences are easier to construct by combining words than by combining letters. Sentences then combine into paragraphs, paragraphs combine into chapters, and, eventually, chapters combine to form a book—highly improbable, but vastly more probable than the chance of arriving at a book by searching the space of possible combinations at the level of letters or words. It was apparent to Merezhkovsky and Kozo-Polyansky that life represents the culmination of a succession of coalitions between simpler organisms, ultimately descended from not-quite-living component parts. Eukaryotic cells are riddled with evidence of symbiotic origins, a view that has been restored to respectability by Lynn Margulis in recent years. But microbiologists arrived too late to witness the symbiotic formation of living cells.

Barricelli enlarged on the theory of cellular symbiogenesis, formulating a more general theory of "symbioorganisms," defined as any "self-reproducing structure constructed by symbiotic association of several self-reproducing entities of any kind."[6] Extending the concept beyond familiar (terrestrial) and unfamiliar (extraterrestrial) chemistries in which populations of self-reproducing molecules might develop by autocatalytic means, Barricelli applied the same logic to self-reproducing patterns of any nature in space or time—such as might be represented by a subset of the 40,960 bits of information, shifting from microsecond to microsecond within the memory of the new machine at the IAS. "The distinction between an evolution experiment performed by numbers in a computer or by nucleotides in a chemical laboratory is a rather subtle one," he observed.[7]

Barricelli saw the IAS computer as a means of introducing self-reproducing structures into an empty universe and observing the results. "The Darwinian idea that evolution takes place by random hereditary changes and selection has from the beginning been handicapped by the fact that no proper test had been found to decide whether such evolution was possible and how it would develop under controlled conditions," he reported in a review of the experiments performed at the IAS. "A test using living organisms in rapid evolution (viruses or bacteria) would have the serious drawback that the causes of adaptation or evolution would be difficult to state unequivocally, and Lamarckian or other kinds of interpretation would be difficult to exclude. However if, instead of using living organisms, one could experiment with entities which, without any doubt could evolve exclusively by 'mutations' and selection, then and only then would a successful evolution experiment give conclusive evidence; the better if the environmental factors also are under control as for example if they are limited to some sort of competition between the entities used."[8]

After forty-three years, Barricelli's experiments appear as archaic as Galileo's first attempt at a telescope—less powerful than half a pair of cheap binoculars—although Galileo's salary was doubled by the Venetian Senate in 1609 as a reward. The two Italians compensated for their primitive instruments with vision that was clear. Barricelli tailored his universe to fit within the limited storage capacity of the IAS computer's forty Williams tubes: a total of one two-hundredth of a megabyte, in the units we use today. Operating systems and programming languages did not yet exist. "People had to essentially program their problems in 'absolute,'" James Pomerene explained, recalling early programming at the IAS, when every single instruction had to be hand-coded to refer to an absolute memory address. "In other words, you had to come to terms with the machine and the machine had to come to terms with you."[9]

Working directly in binary machine instruction code, Barricelli constructed a cyclical universe of 512 cells, each cell occupied by a number (or the absence of a number) encoded by 8 bits. Simple rules that Barricelli referred to as "norms" governed the propagation of numbers (or "genes"), a new generation appearing as if by metamorphosis after the execution of a certain number of cycles by the central arithmetic unit of the machine. These reproduction laws were configured "to make possible the reproduction of a gene only when other different genes are present, thus necessitating symbiosis between different genes."[10] The laws were concise, ordaining only that each number shift to a new location (in the next generation) determined by the location and value of certain genes in the current generation. Genes depended on each other for survival, and cooperation (or parasitism) was rewarded with success. A secondary level of norms (the "mutation rules") governed what to do when two or more different genes collided in one location, the character of these rules proving to have a marked effect on the evolution of the gene universe as a whole. Barricelli played God, on a very small scale.

The empty universe was inoculated with random numbers generated by drawing playing cards from a shuffled deck. Robust and self-reproducing numerical coalitions (patterns loosely interpreted as "organisms") managed to evolve. "We have created a class of numbers which are able to reproduce and to undergo hereditary changes," Barricelli announced. "The conditions for an evolution process according to the principle of Darwin's theory would appear to be present. The numbers which have the greatest survival in the environment . . . will survive. The other numbers will be eliminated little by little. A process of adaptation to the environmental conditions, that is, a process of Darwinian evolution, will take place."[11] Over thousands of generations, Barricelli observed a succession of "biophenomena," such as successful crossing between parent symbioorganisms and cooperative self-repair of damage when digits were removed at random from an individual organism's genes.

The experiments were plagued by problems associated with more familiar forms of life: parasites, natural disasters, and stagnation when there were no environmental challenges or surviving competitors against which organisms could exercise their ability to evolve. To control the parasites that infested the initial series of experiments in 1953, Barricelli instituted modified shift norms that prevented parasitic organisms (especially single-gened parasites) from reproducing more than once per generation, thereby closing a loophole through which they had managed to overwhelm more complex organisms and bring evolution to a halt. "Deprived of the advantage of a more rapid reproduction, the most primitive parasites can hardly compete with the more evolved and better organized species . . . and what in other conditions could be a dangerous one-gene parasite may in this region develop into a harmless or useful symbiotic gene."[12]

Barricelli discovered that evolutionary progress was achieved not so much through chance mutation as through sex. Gene transfers and crossing between numerical organisms were strongly associated with both adaptive and competitive success. "The majority of the new varieties which have shown the ability to expand are a result of crossing-phenomena and not of mutations, although mutations (especially injurious mutations) have been much more frequent than hereditary changes by crossing in the experiments performed."[13] Echoing the question that Samuel Butler had asked seventy years earlier in *Luck, or Cunning?* Barricelli concluded that "mutation and selection alone, however, proved insufficient to explain evolutionary phenomena."[14] He credited symbiogenesis with accelerating the evolutionary process and saw "sexual reproduction [as] the result of an adaptive improvement of the original ability of the genes to change host organisms and recombine."[15] Symbiogenesis leads to parallel processing of genetic code, both within an individual multicellular organism and across the species as a whole. Given that nature allows a plenitude of processors but a limited amount of time, parallel processing allows a more efficient search for those sequences that move the individual, and the species, ahead.

Efficient search is what intelligence is all about. "Even though biologic evolution is based on random mutations, crossing and selection, it is not a blind trial-and-error process," explained Barricelli in a later retrospective of his numerical evolution work. "The hereditary material of all individuals composing a species is organized by a rigorous pattern of hereditary rules into a collective intelligence mechanism whose function is to assure maximum speed and efficiency in the solution of all sorts of new problems . . . and the ability to solve problems is the primary element of intelligence which is used in all intelligence tests. . . . Judging by the achievements in the biological world, that is quite intelligent indeed."[16]

A century after *On the Origin of Species* pitted Charles Darwin and Thomas Huxley against Bishop Wilberforce, there was still no room for compromise between the trial and error of Darwin's natural selection and the supernatural intelligence of a theological argument from design. Samuel Butler's discredited claims of species-level intelligence—neither the chance success of a blind watchmaker nor the predetermined plan of an all-knowing God—were reintroduced by Barricelli, who claimed to detect faint traces of this intelligence in the behavior of pure, self-reproducing numbers, just as viruses were first detected by biologists examining fluids from which they had filtered out all previously identified living forms.

The evolution of digital symbioorganisms took less time to happen than to describe. "Even in the very limited memory of a high speed computer a large number of symbioorganisms can arise by chance in a few seconds," Barricelli reported. "It is only a matter of minutes before all the biophenomena described can be observed."[17] The digital universe had to be delicately adjusted so that evolutionary processes were not immobilized by dead ends. Scattered among the foothills of the evolutionary fitness landscape were local maxima from which "it is impossible to change only one gene without getting weaker organisms." In a closed universe inhabited by simple organisms, the only escape to higher ground was by exchanging genes with different organisms or by local shifting of the rules. "Only replacements of at least two genes can lead from a relative maximum of fitness to another organism with greater vitality,"[18] noted Barricelli, who found that the best solution to these problems (besides the invention of sex) was to build a degree of diversity into the universe itself.

"The Princeton experiments were continued for more than 5,000 generations using universes of 512 numbers," Barricelli reported. "Moreover, the actual size of the universe was usually increased far beyond 512 numbers by running several parallel experiments with regular interchanging of several (50 to 100) consecutive numbers between two universes. . . . Within a few hundred generations a single primitive variety of symbioorganism invaded the whole universe. After that stage was reached no collisions leading to new mutations occurred and no evolution was possible. The universe had reached a stage of 'organized homogeneity' which would remain unchanged for any number of following generations. . . . In many instances a new mutation rule would lead to a complete disorganization of the whole universe, apparently due to the death by starvation of a parasite, which in this case was the last surviving organism. . . . Homogeneity problems were eventually overcome by using different mutation rules in different sections of each universe. Also slight modifications of the reproduction rule were used in different universes to create different types of environment . . . by running several parallel experiments and by exchanging segments between two universes every 200 or 500 generations it was possible to break homogeneity whenever it developed in one of the universes."[19]

As Alan Turing had blurred the distinction between intelligence and nonintelligence by means of his universal machine, so Barricelli's numerical symbioorganisms blurred the distinction between living and nonliving things. Barricelli cautioned his audience against "the temptation to attribute to the numerical symbioorganisms a little too many of the properties of living beings," and warned that "the author takes no responsibility for inferences and interpretations which are not rigorous consequences of the facts presented."[20] He stressed that although numerical symbioorganisms and known terrestrial life-forms exhibited parallels in evolutionary behavior, this did not imply that numerical symbioorganisms were alive. "Are they the beginning of, or some sort of, foreign life forms? Are they only models?" he asked. "They are not models, not any more than living organisms are models. They are a particular class of self-reproducing structures already defined." As to whether they are living, "it does not make sense to ask whether symbioorganisms are living as long as no clear-cut definition of 'living' has been given."[21] A clear-cut definition of "living" remains elusive to this day.

Barricelli's numerical organisms were like tropical fish in an aquarium, confined to an ornamental fragment of a foreign ecosystem, sealed behind the glass face of a Williams tube. The perforated cards that provided the only lasting evidence of their existence were lifeless imprints, skeletons preserved for study and display. The numerical organisms consisted of genotype alone and were far, far, simpler than even the most primitive viruses found in living cells (or computer systems) today. Barricelli knew that "something more is needed to understand the formation of organs and properties with a complexity comparable to those of living organisms. No matter how many mutations occur, the numbers . . . will never become anything more complex than plain numbers."[22] Symbiogenesis—the forging of coalitions leading to higher levels of complexity—was the key to evolutionary success, but success in a closed, artificial universe has only fleeting meaning in our own. Translation into a more tangible phenotype (the interpretation or execution, whether by physical chemistry or other means, of the organism's genetic code) was required to establish a presence in our universe, if Barricelli's numerical symbioorganisms were to become more than laboratory curiosities, here one microsecond and gone the next.

Barricelli wondered "whether it would be possible to select symbioorganisms able to perform a specific task assigned to them. The task may be any operation permitting a measure of the performance reached by the symbioorganisms involved; for example, the task may consist in deciding the moves in a game being played against a human or against another symbioorganism."[23] In a later series of experiments (performed on an IBM 704 computer at the AEC computing laboratory at New York University in 1959 and at Brookhaven National Laboratory in 1960) Barricelli evolved a class of numerical organisms that learned to play a simple but nontrivial game called "Tac-Tix," played on a 6-by-6 board and invented by Piet Hein. The experiment was configured so as to relate game performance to reproductive success. "With present speed, it may take 10,000 generations (about 80 machine hours on the IBM 704 . . .) to reach an average game quality higher than 1," Barricelli estimated, this being the quality expected of a rank human beginner playing for the first few times.[24] In 1963, using the large Atlas computer at Manchester University, this objective was achieved for a short time, but without further improvement, a limitation that Barricelli attributed to "the severe restrictions . . . concerning the number of instructions and machine time the symbioorganisms were allowed to use."[25]

In contrast to the IAS experiments, in which the numerical symbioorganisms consisted solely of genetic code, the Tac-Tix experiments led to "the formation of non-genetic numerical patterns characteristic for each symbioorganism. Such numerical patterns may present unlimited possibilities for developing structures and organs of any kind to perform the tasks for which they are designed."[26] A numerical phenotype had taken form. This phenotype was interpreted as moves in a board game, via a limited alphabet of machine instructions to which the gene sequence was mapped, just as sequences of nucleotides code for an alphabet of amino acids in translating proteins from DNA. "Perhaps the closest analogy to the protein molecule in our numeric symbioorganisms would be a subroutine which is part of the symbioorganism's game strategy program, and whose instructions, stored in the machine memory, are specified by the numbers of which the symbioorganism is composed," Barricelli explained.[27] In coding for valid instructions at the level of phenotype rather than genotype, evolutionary search is much more likely to lead to meaningful sequences, for the same reason that a meaningful sentence is far more likely to be evolved by choosing words out of a dictionary than by choosing letters out of a hat.

A purely numerical sequence could in principle (and in time) evolve to be translated, through any number of intermediary languages, into anything else. "Given a chance to act on a set of pawns or toy bricks of some sort the symbioorganisms will 'learn' how to operate them in a way which increases their chance for survival," Barricelli explained. "This tendency to act on any thing which can have importance for survival is the key to the understanding of the formation of complex instruments and organs and the ultimate development of a whole body of somatic or non-genetic structures."[28] Once the concept of translation from genotype to phenotype is given form, Darwinian evolution picks up speed—not just the evolution of organisms, but the evolution of the genetic language and translation system that provide the flexibility and redundancy to survive in a noisy, unpredictable world. A successful interpretive language not only tolerates ambiguity, it takes advantage of it. "It is almost too easy to imagine possible uses for phenotype structures—because the specification for an effective phenotype is so sloppy," wrote A. G. Cairns-Smith in his *Seven Clues to the Origin of Life.* "A phenotype has to make life easier or less dangerous for the genes that (in part) brought it into existence. There are no rules laid down as to how this should be done."[29]

Barricelli's pronouncements had a vaguely foreboding, Butlerish air about them, despite the disclaimer about confusing "life-like" with "alive." Samuel Butler had warned that Darwin's irresistible logic applied not only to the kingdom of nature but to the kingdom of machines; Nils Barricelli now demonstrated that it was the kingdom of numbers that held the key to that "Great First Cause" of Erasmus Darwin's "one living filament," whether encoded as strings of nucleotides or as strings of electronic bits. Barricelli saw that electronic digital computers heralded an unprecedented change of evolutionary pace, just as Butler had seen evolution quickened by the age of steam.

Barricelli, like Samuel Butler, was a nonconformist whose contributions were obscured by lingering arguments with the authorities of his time. After receiving a degree in mathematics and physics at the University of Rome in 1936, he emigrated, out of opposition to the Fascists, to Norway in 1938. He prepared his doctoral thesis (on climate variations) during the war, submitting the thesis in 1946. "However it was 500 pages long, and was found to be too long to print," remarked Barricelli's former student and fellow mathematician Tor Gulliksen. "He did not agree to cut it to an acceptable size, and chose instead not to obtain the doctoral degree!"[30] Gulliksen worked as Barricelli's assistant for the summer of 1962 and 1963, using the Atlas computer at Manchester, the most powerful computer of its time, in an attempt to evolve numerical symbioorganisms for playing chess.

After a series of visiting appointments in virus genetics and theoretical biology, Barricelli returned to the University of Oslo in

1969, where he remained a guest of the Mathematical Research Institute until his death—"but without a salary," said Gulliksen. "He would rather keep his complete freedom as a researcher than enter a permanent position at the University."[31] Barricelli believed that an inconsistency was concealed within Kurt Gödel's 1931 incompleteness results, a suspicion that alienated him from the mathematical establishment and led to the publication of a series of research papers on this and other subjects at his own expense. "He believed that every mathematical statement could either be proved or disproved. He insisted that Gödel's proof was faulty," said Simen Gaure, who assisted with Barricelli's proof generator project in 1983–1985. Gaure was hired ("He paid us directly out of his wallet, fairly good pay it was too, at least for students") after a selection process that required searching for a deeply hidden flaw in a sample proof. "Those who could point out the flaw were accepted as not yet ruined by mathematical education," reported Gaure. Barricelli intended "to actually build a machine which could prove or disprove any statement of arithmetic and projective geometry. He developed something he called 'B-mathematics,' which was a logic language particularly well suited for this task. . . . The proof constructor programs were written in Fortran and Simula on a DEC system 10 computer. I once asked him what the 'B' in 'B-math' was. He answered that he hadn't decided on that; it could be 'Boolean,' or it could be 'Barricelli,' or something else, he said."[32]

Barricelli and the B-mathematics language died together in January 1993. "Languages as well as species evolve by mutation, crossing and selection," Barricelli had noted in 1966. "A language cannot survive or propagate unless there are humans or whatever species is using the language. But that is a property it has in common with many symbionts and parasites. One may consider a language as a symbiont of *Homo sapiens* of a different nature than the nucleic acid–protein combinations we are used to consider as living organisms."[33] Barricelli's language did not outlive its host. But his numerical symbioorganisms may still be lying dormant in a deck of punched cards somewhere, surviving long after magnetically stored data have decayed. "He insisted on using punch cards, even when everybody had computer screens," Gaure remembered. "He gave two reasons for this: when you sit in front of a screen your ability to think clearly declines because you're distracted by irrelevancies, and when you store your data on magnetic media you can't be sure they're there permanently, you actually don't know where they are at all."[34]

Barricelli's use of biological terminology to describe self-reproducing code fragments is reminiscent of early pronouncements about artificial intelligence, when machines that processed information with less intelligence than a pocket calculator were referred to as machines that think. A relic from the age of vacuum tubes and giant brains, Barricelli's IAS experiments strike the modern reader as naïve—until you stop and reflect that numerical symbioorganisms have, in less than fifty years, proliferated explosively, deeply infiltrating the workings of life on earth. With our cooperation as symbiotic hosts, self-reproducing numbers are managing (Barricelli would say learning) to exercise increasingly detailed and far-reaching control over the conditions in our universe that are helping to make life more comfortable in theirs. Are the predictions of Samuel Butler and Nils Barricelli turning out to be correct?

"Since computer time and memory still is a limiting factor, the non-genetic patterns of each numeric symbioorganism are constructed only when they are needed and are removed from the memory as soon as they have performed their task," explained Barricelli, describing the Tac-Tix–playing organisms of 1959. He might as well have been describing that class of numerical symbioorganisms—computer software—that we execute and terminate from moment to moment today. "This situation is in some respects comparable to the one which would arise among living beings if the genetic material got into the habit of creating a body or a somatic structure only when a situation arises which requires the performance of a specific task (for instance a fight with another organism), and assuming that the body would be disintegrated as soon as its objective had been fulfilled."[35]

The precursors of symbiogenesis in the von Neumann universe were order codes, conceived (in the Burks–Goldstine–von Neumann reports) before the digital matrix that was to support their existence had even taken physical form. Order codes constituted a fundamental alphabet that diversified in association with the proliferation of different hosts. In time, successful and error-free sequences of order codes formed into subroutines—the elementary units common to all programs—just as a common repertoire of nucleotides is composed into strings of DNA. Subroutines became organized into an expanding hierarchy of languages, which then influenced the computational atmosphere as pervasively as the oxygen released by early microbes influenced the subsequent course of life.

By the 1960s complex numerical symbioorganisms known as operating systems had evolved, bringing with them entire ecologies of symbionts, parasites, and coevolving hosts. The most successful operating systems, such as OS/360, MS-DOS, and UNIX, succeeded in transforming and expanding the digital universe to better propagate themselves. It took five thousand programmer-years of effort to write and debug the OS/360 code; the parasites and symbionts sprouted up overnight. There was strength in numbers. "The success of some programming systems depended on the number of machines they would run on," commented John Backus, principal author of Fortran, a language that has had a long and fruitful symbiosis with many hosts.[36] The success of the machines depended in turn on their ability to support the successful languages; those that clung to dead languages or moribund operating systems became extinct.

The computational ecology grew by leaps and bounds. In 1954, IBM's model 650 computer shipped with 6,000 lines of code; the first release of OS/360 in 1966 totaled just under 400,000 instructions, expanding to 2 million instructions by the early 1970s. The total of all system software provided by the major computer manufacturers reached 1 million lines of code by 1959, and 100 million by 1972. The amount of random-access memory in use worldwide, costing an average $4.00 per byte, reached a total of 1,000 megabytes in 1966, and a total of 10,000 megabytes, at an average cost of $1.20 per byte, in 1971. Annual sales of punched cards by U.S. manufacturers exceeded 200 billion cards (or 500,000 tons) in 1967, after which the number of cards began to decline in favor of magnetic tapes and disks.[37]

In the 1970s, with the introduction of the microprocessor, a second stage of this revolution was launched. The replication of processors thousands and millions at a time led to the growth of new forms of numerical symbioorganisms, just as the advent of metazoans sparked a series of developments culminating in an explosion of new life-forms six hundred million years ago. New species of numerical symbioorganisms began to appear, reproduce, and become extinct at a rate governed by the exchange of floppy disks rather than the frequency of new generations of mainframes at IBM. Code was written, copied, combined, borrowed, and stolen among software producers as freely as in a primordial soup of living but only vaguely differentiated cells. Anyone who put together some code that could be executed as a useful process—like Dan Fylstra's Visicalc in 1979—was in for a wild ride. Businesses sprouted like mushrooms, supported by the digital mycelium underneath. Corporations came and went, but successful code lived on.

Twenty years later, fueled by an epidemic of packet-switching protocols, a particularly virulent strain of symbiotic code, the neo-Cambrian explosion entered a third and yet more volatile phase. Now able to replicate at the speed of light instead of at the speed of circulating floppy disks, numerical symbioorganisms began competing not only for memory and CPU cycles within their local hosts but within a multitude of hosts at a single time. Successful code is now executed in millions of places at once, just as a successful genotype is expressed within each of an organism's many cells. The possibilities of complex, multicellular digital organisms are only beginning to be explored.

The introduction of distributed object-oriented programming languages (metalanguages, such as Java, that allow symbiogenesis to transcend the proprietary divisions between lower-level languages in use by different hosts) is enabling numerical symbioorganisms to roam, reproduce, and execute freely across the computational universe as a whole. Through the same hierarchical evolution by which order codes were organized into subroutines and subroutines into programs, objects, being midlevel conglomerations of code, will form higher-level structures distributed across the net. Object-oriented programming languages were first introduced some years ago with a big splash that turned out to be a flop. But what failed to thrive on the desktop may behave entirely differently on the Internet. Nils Barricelli, in 1985, drew a parallel between higher-level object-oriented languages and the metalanguages used in cellular communication, but he put the analogy the other way: "If humans, instead of transmitting to each other reprints and complicated explanations, developed the habit of transmitting computer programs allowing a computer-directed factory to construct the machine needed for a particular purpose, that would be the closest analogue to the communication methods among cells of various species."[38]

But aren't these analogies deeply flawed? Software is designed, engineered, and reproduced by human beings; programs are not independently self-reproducing organisms selected by an impartial environment from the random variation that drives other evolutionary processes that we characterize as alive. The analogy, however, is valid, because the analog of software in the living world is not a self-reproducing organism, but a self-replicating molecule of DNA. Self-replication and self-reproduction have often been confused. Biological organisms, even single-celled organisms, do not replicate themselves; they host the replication of genetic sequences that assist in reproducing an approximate likeness of themselves. For all but the lowest organisms, there is a lengthy, recursive sequence of nested programs to unfold. An elaborate self-extracting process restores entire directories of compressed genetic programs and reconstructs increasingly complicated levels of hardware on which the operating system runs. That most software is parasitic (or symbiotic) in its dependence on a host metabolism, rather than freely self-replicating, strengthens rather than weakens the analogies with life.

The randomness underlying evolutionary processes has also been overplayed. Many seemingly haphazard genetic processes are revealing themselves to be less random than once thought. By means of higher-level languages, grammars, and error-correcting codes, much of the randomness is removed before a hereditary message is submitted for execution as another cell. A certain collective intelligence adheres to the web of relationships among genetic regulators and operators, a vague and faintly distributed unconscious memory that raises Samuel Butler's ghost. What randomness *does* contribute to evolutionary processes is a small but measurable element of noise. By definition, a Darwinian process has an element of randomness—but it does not have to be a game of chance. By almost any standard, the software industry qualifies as a Darwinian process—from the generation of software (you combine existing code segments, then execute and weed out the bugs) to the management of the industry as a whole: eat, be eaten, or merge.

No one can say what contribution randomness has made to software development so far. Most programs have grown so complex and convoluted that no human being knows where all the code came from or even what some of it actually does. Programmers long ago gave up hope of being able to predict in advance whether a given body of code will work as planned. "It was on one of my journeys between the EDSAC room and the punching equipment," recalled Maurice Wilkes of one particular day of program testing at Cambridge in 1949, "that 'hesitating at the angle of the stairs' the realization came over me with full force that a good part of the remainder of my life was going to be spent in finding errors in my own programs."[39] The software industry has kept track of harmful bugs since the beginning—but there is no way to keep track of the accidents and coincidences that have accumulated slight improvements along the way.

In 1953, Nils Barricelli observed a digital universe in the process of being born. There was only a fraction of a megabyte of random-access memory on planet Earth, and only part of it was working at any given time. "The limited capacity of even the largest calculating machines makes it impossible to operate with more than a few thousand genes at a time instead of the thousands of billions of genes and organisms with which nature operates," he wrote in 1957. "This makes it impossible to develop anything more than extremely primitive symbioorganisms even if the most suitable laws of reproduction are chosen."[40] Not so today. Barricelli's universe has expanded explosively, providing numerical organisms with inexhaustible frontiers on which to grow.

Barricelli's contributions have largely been forgotten. Few later writers have searched the literature of the 1950s for signs of artificial life. It is hard to believe that such experiments could be performed with the equipment available in 1953. Von Neumann's outline of a formal theory of self-reproducing automata, developed in the 1950s but published only in 1966, is regarded as the precursor of the field that would develop, decades later, into the study of artificial life. Von Neumann took a theoretical, not an experimental, approach. Although Barricelli came to the Institute at von Neumann's invitation, the same caution with which von Neumann viewed speculations about artificial intelligence was applied to Barricelli's suggestion that numerical symbioorganisms had started to evolve. Times have changed. "A-life" is now a legitimate discipline, and researchers need not preface their papers with disclaimers lest any reviewer think they are suggesting that numerical organisms evolved within a computer might be alive.

Among the most creative of Nils Barricelli's successors is evolutionary biologist Thomas Ray. After a decade of fieldwork in the Central American rain forest ("The rainforest is like a huge cathedral, but the entire structure is alive") Ray grew impatient with the pace of evolution—too slow to observe except by studying the past. "The greatest obstacles to understanding evolution have been that we have had only a single example of evolution available for study (life on earth), and that in this example, evolution is played out over time spans which are very large compared to a scientific career," wrote Ray in 1993, explaining his decision to experiment with evolution in a faster form.[41] The result was Tierra, a stripped-down, 5-bit instruction-code virtual computer that can be embedded within any species of host. Tierra (Spanish for "earth") is designed to provide a fertile and forgiving environment in which self-replicating digital organisms can evolve. Tierra borrows heavily from nature, for instance by allowing the organisms to locate each other using complementary templates, as in molecular biology, rather than by numerical address. Ray expected to fiddle around fruitlessly with his new system, but, as he recounted, "my plans were radically altered by what actually happened on the night of January 3, 1990, the first time that my self-replicating program ran on my virtual computer, without crashing the real computer that it was emulated on. . . . All hell broke loose. The power of evolution had been unleashed inside the machine, but accelerated to . . . megahertz speeds."[42]

"My research program was suddenly converted from one of design, to one of observation," reported Ray. "I was back in a jungle describing . . . an alien universe, based on a physics and chemistry totally different than the life forms I knew and loved. Yet forms and processes appeared that were somehow recognizable to the trained eye of a naturalist."[43] The Tierran organisms were limited to 80 bytes apiece, yet far-reaching speculations were inspired by Ray's glimpse into their world. "While these new living systems are still so young that they remain in their primordial state, it appears that they have embarked on the same kind of journey taken by life on earth, and presumably have the potential to evolve levels of complexity that could lead to sentient and eventually intelligent beings," he wrote in 1993.[44]

Ray came to the same conclusions as had Barricelli concerning the conditions his numerical organisms required to continue to evolve. Communities of coevolving digital species need large, complex spaces in which to grow. The challenge of semipermeable boundaries between diverse and changing habitats is reflected in increasingly diverse and complex organisms. Working with Kurt Thearling of Thinking Machines, Inc., Ray used a massively parallel CM-5 Connection Machine to develop an archipelago of Tierran nodes, each running a slightly different Tierran soup, with occasional migration of creatures between different nodes, similar to the multiple 512-byte universes Barricelli had constructed at the IAS. But even the universe within the Connection Machine was only a local node within the open universe that awaits. "Due to its size, topological complexity, and dynamically changing form and conditions, the global network of computers is the ideal habitat for the evolution of complex digital organisms," concluded Ray.[45]

To conduct a full-scale experiment, Ray proposed (and has now begun to construct) a "biodiversity reserve for digital organisms" distributed across the global net. "Participating nodes will run a network version of Tierra as a low-priority background process, creating a virtual Tierra sub-net embedded within the real net . . . there will be selective pressures for organisms to migrate around the globe in a daily cycle, to keep on the dark side of the planet, and also to develop sensory capabilities for assessing deviations from the expected patterns of energy availability, and skills at navigating the net."[46]

Having experienced firsthand the difficulty of convincing politicians and resource managers of the value of biological reserves, Ray pitched his proposal with the same arguments that apply to preserving the tropical rain forest as a reservoir of irreplaceable genetic code and yet-to-be-discovered drugs. The Tierran reserve is envisioned as a cooperative laboratory for evolving commercially harvestable software of a variety and complexity beyond what we could ever hope to engineer. "This software will be 'wild,' living free in the digital biodiversity reserve," proposed Ray. "In order to reap the rewards, and create useful applications, we will need to domesticate some of the wild digital organisms, much as our ancestors began domesticating the ancestors of dogs and corn thousands of years ago."[47] The potentially useful products might range from simple code fragments, equivalent to drugs discovered in rain-forest plants, to entire digital organisms, which Ray imagines as evolving into digital counterparts of the one-celled workhorses that give us our daily bread and cheese and beer—and eventually into higher forms of digital life. "It is imagined," wrote Ray, "that individual digital organisms will be multi-celled and that the cells that constitute an individual will be dispersed over the net. . . . If there are some massively parallel machines participating in the virtual net, digital organisms may choose to deploy their central nervous systems on these arrays."[48]

Ray's proposal leaves one wondering what will keep these wild organisms from making an escape—a scenario labeled Jurassic Park. The answer is that Tierran organisms can survive only within the universe of virtual machines in which they evolved. Outside this special environment they are only data, no more viable than any other data that reside on the machines with which we work and play. Nonetheless, Ray advised that "freely evolving autonomous artificial entities should be seen as potentially dangerous to organic life, and should always be confined by some kind of containment facility, at least until their real potential is well understood. . . . Evolution remains a self-interested process, and even the interests of confined digital organisms may conflict with our own."[49]

In March 1995, Ray convened a symposium on the network implementation of Tierra at which the security issue was addressed. "There was unanimous agreement that this Terminator 2/Jurassic Park scenario is not a security issue," reported Ray. "Nobody at the workshop considered this to be a realistic possibility, because the digital organisms are securely confined within the virtual net of virtual machines." What concerned the security experts, led by Tsutomu Shimomura, was the possibility of bad people breaking *into* the Tierran system, rather than the possibility of bad organisms breaking *out*. "Tierra is a general purpose computer . . . and a large network implementation would be a very large general purpose computer, perhaps the largest in the world." This resource might conceivably be expropriated by outside users, which "would be equivalent to cutting down the rain forest to plant bananas," according to Ray. If Tierran processes were cultivated for illicit purposes, say, for breaking codes, "this would be like cutting down the rain forest to plant cocaine."[50]

Left unsaid in these discussions is the extent to which the computational universe has already become a jungle, teeming with freely evolving life. Whether the powers of digital evolution can be kept in reserve, without running wild, is a question that depends on how you define wild, just as the question of artificial intelligence depends on how you define intelligence, and the question of artificial life depends on how you define alive. The whole point of the Tierran experiment is to adopt successful Tierran code, carefully neutered, into software running outside the boundaries of the reserve. No matter how this arrangement is described, it means that the results of freely evolving processes will be introduced into the digital universe as a whole. Harmful results will be edited out (assuming there are no mistakes), but that's the way distribution of code in biology has always worked. The effect will be to greatly speed up the evolution of digital organisms, whether the underlying mechanism is kept within a securely guarded reserve or not.

All indications point to a convergence between the Tierran universe as envisioned by Tom Ray and the computational universe as a whole. Platform-independent languages such as Java and its siblings are moving in the direction of universal code that runs on a universal virtual machine, spawned on the host processor of the moment, with a subtle shift in computational identity as objects are referred to more by process than by place. Such a language falls somewhere between the mechanical notation of Charles Babbage, able to describe and translate the logical function of any conceivable machine, and the language of DNA, able to encode the construction of proteins in a form that can be translated by diverse forms of life. Data as well as programs are becoming migratory, and objects that are used frequently find themselves mirrored in many places at once. As successful processes proliferate throughout the network, a variety of schemes for addressing objects by template rather than by location are likely to take form. The result will be a profoundly Tierran shift in the computational landscape—but with the direct links to the workings of our own universe that are essential if digital organisms that actually go out and do things are to continue to evolve.

"Given enough time in a sufficiently varied universe," predicted Nils Barricelli, "the numeric symbioorganisms might be able to improve considerably their technique in the use of evolutionary processes."[51] Cultivating a genetic diversity reserve on the scale of a network implementation of Tierra would be a clever improvement indeed. Not so clever, however, that biology has not had the time to get there first. DNA-based life maintains a global reserve of microorganisms, able to adapt to changing circumstances at a much faster pace than can more complex organisms—which occasionally suffer from, but ultimately reap the benefits of, new acquisitions made by this public library of genetic code.

In later years, Barricelli continued to apply the perspective gained through his IAS experiments to the puzzle of explaining the origins and early evolution of life. "The first language and the first technology on Earth was not created by humans. It was created by primordial RNA molecules almost 4 billion years ago," he wrote in 1986. "Is there any possibility that an evolution process with the potentiality of leading to comparable results could be started in the memory of a computing machine and carried on to a stage giving fundamental information on the nature of life?" He endeavored "to obtain as much information as possible about the way in which the genetic language of the living organisms populating our planet (terrestrial life forms) originated and evolved."[52] Barricelli viewed the genetic code "as a language used by primordial 'collector societies' of t[ransfer]RNA molecules . . . specialized in the collection of amino acids and possibly other molecular objects, as a means to organize the delivery of collected material." He drew analogies between this language and the languages used by other collector societies, such as social insects, but warned that "trying to use the ant and bee languages as an explanation of the origin of the genetic code would be a gross misunderstanding."[53] Languages are, however, the key to evolving increasingly complex, self-reproducing structures through the cooperation of simpler component parts.

According to Simen Gaure, Nils Barricelli "balanced on a thin line between being truly original and being a crank." Most cranks turn out to be cranks; a few cranks turn out to be right. "The scientific community needs a couple of Barricellis each century," added Gaure. As Barricelli's century draws to a close, the distinctions between A-life (represented by strings of electronic bits) and B-life (represented by strings of nucleotides) are being traversed by the first traces of a language that comprehends them both. Does this represent the gene's learning to manipulate the power of the bit, or does it represent the bit's learning to manipulate the power of the gene? As algae and fungi became lichen, the answer will be both. And it is the business of symbiogenesis to bring such coalitions to life.

In the mathematical universe of Kurt Gödel, all of creation—mathematical objects, theorems, concepts, and ideas—can be identified by individual Gödel numbers, establishing a numerical bureaucracy that in Alan Turing's computational universe was extended to include machines and even organisms, specified by a "state of mind" and a "description number" of some particular (or equivalent) Turing machine. Any Turing machine, and any state of mind, can be encoded, however tediously, as a sequence of 0s and 1s. Gödel and Turing demonstrated how this universe could be populated by an infinite succession of increasingly powerful languages, thereby proving that we live in a mathematically open universe whose boundaries will never be closed. In the 1930s, no one imagined this formalized, discrete-state universe ever taking actual physical form. "When I was a student, even the topologists regarded mathematical logicians as living in outer space," recalled Martin Davis, who hand-coded the first theorem-proving program in 1954 at the IAS. "Today ... one can walk into a shop and ask for a 'logic probe.' "[54]

Where logic led, electronics followed. Thanks to Gödel, Turing, and colleagues, the proof was there from the beginning that a digital universe would be an open universe in which mathematical structures of unbounded complexity, intellect, meaning, and even beauty might freely grow. There is no limit, in mathematics or in physics, to how far and how fast Barricelli's numerical symbioorganisms will be able to evolve. "This process will be more expeditious than evolution," Alan Turing predicted in 1950. "The survival of the fittest is a slow method for measuring advantages. The experimenter, by the exercise of intelligence, should be able to speed it up."[55]

8

ON DISTRIBUTED COMMUNICATIONS

Real wires take up room.

—W. DANIEL HILLIS[1]

The latest developments in telecommunications are all-optical data networks. So were the first. The recorded history of high-speed optical data transmission began with the fall of Troy to the Mycenaean army, allegedly in 1184 B.C. Across the Aegean in Mycenae, as legend has it, Clytaemnestra awaited news from her husband, Agamemnon, absent for ten years in the course of the siege. When Troy was taken, a prearranged signal was relayed overnight to Mycenae, via a line of fire beacons, a distance of some 375 miles, much of it over sea. The tragedy of *Agamemnon,* chronicled by Aeschylus (525–456 B.C.), opens with Clytaemnestra receiving news of the signal from her watchman while the chorus asks: "And what messenger is there that could arrive with such speed as this?"

Clytaemnestra answers: "Hephaistos [God of fire], sending forth from Ida a bright radiance. And beacon ever sent beacon hither by means of the courier fire: Ida (sent it) to the rock of Hermes in Lemnos; and a huge torch from the island was taken over in the third place by Zeus' peak of Athos; and paying more than what was due, so as to skim the back of the sea . . . transmitting, like a sun, its golden radiance to the look-out of Makistos. And he, not dallying nor heedlessly overcome by sleep, did not neglect his share in the messenger's duty, and afar, over the streams of Euripus, the beacon's light gave the watchers of Messapion the sign of its arrival. They kindled an answering flare and sent the tidings onward, by setting fire to a stack of aged heath. And the vigorous torch, not yet growing dim, leaped, like the shining moon, over the plain of Asopus to the rock of Kithairon and there waked a new relay of the sender fire. And the

131

far-sent light . . . shot down over the Gorgon-eyed lake and reaching the mountain of the roaming goats. . . . And they with stintless might kindled and sent on a great beard of flame, and it passed beyond the promontory that looks down on the Saronic straits, blazing onward, and shot down when it reached the Arachnaean peak, the watch-post that is neighbour to our city; and then it shot down here to the house of the Atridae, this light, the genuine offspring of its ancestor, the fire from Mount Ida . . . transmitted to me by my husband from Troy."[2]

The link between Troy and Mycenae was a one-way, one-time, and one-bit channel, encoded as follows: no signal meant Troy belonged to the Trojans; a visible signal meant Troy belonged to the Greeks. Communications engineers have been improving the bandwidth ever since. Suffering a fate that still afflicts brief messages after three thousand years, Clytaemnestra's message acquired a header—a cumulative listing of gateways that handled the message along the way—longer than the message she received.

A thousand years later, the Greek historian Polybius (ca. 200–118 B.C.) reported how torch telegraphy had been improved. "The most recent method, devised by Cleoxenus and Democleitus and perfected by myself, is quite definite and capable of dispatching with accuracy every kind of urgent message." The key was to "take the alphabet and divide it into five parts, each consisting of five letters." These five divisions of the twenty-four-letter Greek alphabet were inscribed on five tablets. The transmitting station, after signaling and receiving acknowledgment of the beginning of a transmission by raising two torches, "will now raise the first set of torches on the left side indicating which tablet is to be consulted, i.e., one torch if it is the first, two if it is the second, and so on. Next he will raise the second set on the right on the same principle to indicate what letter of the alphabet the receiver should write down."[3] It would be another two thousand years before modern telegraphy instituted a digital coding of the alphabet as concise and unambiguous as this.

In the seventeenth century, the advantages of a 5-bit alphabetic code were explained by John Wilkins (1614–1672) in a treatise on cryptography, binary coding, and telecommunications titled *Mercury, or the Secret and Swift messenger: Shewing, How a Man may with Privacy and Speed communicate his Thoughts to a Friend at any distance*, published in 1641. Wilkins, who founded the "Experimentall Philosophicall Clubbe" at Oxford in 1649, married Oliver Cromwell's sister in 1656. He was appointed master of Trinity College, Cambridge, in 1659, first secretary of the Royal Society in 1662, and bishop of Chester in 1668.

Wilkins noted that "two letters of the alphabet, being transposed through five places, will yield thirty-two differences, and so will more than serve for the foure and twenty letters unto which they may be thus applyed."[4] After showing how this 5-bit binary coding could be conveyed by torch signals and enciphered in numerous ingenious ways, Wilkins described how to transmit alphabetic text as a sequence of binary acoustic signals, anticipating the modern use of binary coding to transmit text-based intelligence, and to feed the acoustic delay-line storage that gave the stored-program computer industry its start. "It is requisite, that there be two Bels of different notes, or some such other audible and loud sounds, which we may command at pleasure," wrote Wilkins. "By the various soundes of these (according to the former table) a man may easily espresse any letter and so consequently any sense."[5]

Two distinct functions are required of a successful telegraphic code: the encoding of protocols that regulate the process of communication, and the encoding of symbols representing the message to be conveyed. Meaning—in telegraphy as in biology—is encoded hierarchically: first by mapping elementary symbols to some kind of alphabet, then by mapping this alphabet to words, phrases, standard messages, and anything else that can be expressed by brief sequences of code. Higher levels of meaning arise as further layers of interpretation evolve. Protocols, or handshaking, initiate the beginning and end of a transmission and may be used to coordinate error correction and flow control. As Gerard Holzmann and Björn Pehrson observed in their definitive *Early History of Data Networks*, "Some type of protocol has to be established between sender and receiver to deal minimally with the basic problems of synchronization ('after you,' 'no, after you!'), visibility ('repeat please'), and transmission speed ('not so fast!')."[6]

Telecommunications systems have appeared, disappeared, and reappeared across the centuries: fire beacons, heliographs, and primitive forms of semaphore based on hanging or waving anything from flags to lanterns in the air. When the Spanish armada entered the English Channel in July 1588, a network of fire beacons raised the alarm, cradling the newborn Thomas Hobbes with fear. The invention of the telescope in the early seventeenth century extended the distance between relay stations and allowed more complex symbols to be distinguished. The feasibility of a "method of discoursing at a Distance, not by Sound, but by Sight" was addressed by Robert Hooke in a lecture, "Shewing a Way how to communicate one's Mind at great Distances," delivered to the Royal Society on 21 May 1684. Having advanced the optical instruments of his day, Hooke showed that "'tis possible to convey Intelligence from any one high and eminent Place, to any other that lies in Sight of it, tho' 30 or 40 Miles distant, in as short a Time almost, as a Man can write what he would have sent, and as suddenly to receive an Answer as he that receives it hath a Mind to return it. . . . Nay, by the Help of three, four, or more such eminent Places, visible to each other . . . 'tis possible to convey Intelligence, almost in a Moment, to twice, thrice, or more Times that Distance, with as great a Certainty as by Writing."[7]

Robert Hooke (1635–1703) was a brilliant but difficult character whose "temper was Melancholy, Mistrustful and Jealous, which more increas'd upon him with his Years."[8] Possessed of "indefatigable Genius," his creative output was astounding, despite ill humor and ill health. "He is of prodigious inventive head," reported his contemporary John Aubrey, adding that "now when I have sayd his Inventive faculty is so great, you cannot imagine his Memory to be excellent, for they are like two Bucketts, as one goes up, the other goes downe. He is certainly the greatest Mechanick this day in the world."[9] In 1655, Hooke was appointed assistant to Robert Boyle, executing the construction of Boyle's air pump or pneumatic engine with an ingenuity that descended directly, via Thomas Newcomen's atmospheric engine, to the steam engines of the Industrial Revolution and thence to all internal combustion engines in circulation today. After a meeting of the Royal Society on 15 February 1664 (adjourned to the Crown Tavern until ten o'clock that night), Samuel Pepys noted in his diary that "Above all, Mr. Boyle was at the meeting, and above him Mr. Hooke, who is the most, and promises the least, of any man in the world that ever I saw."[10]

In November 1662, Newton, Boyle, and others on the Royal Society's Council established the position of curator of experiments "and order'd that Mr Hooke should come and sit among them, and both bring in every Day three or four of his own Experiments, and take care of such others as should be recommended to him."[11] The ensuing thirty-six years of experimental research were interrupted only by a brief recess during the worst months of the plague in 1665, followed by a period of distraction after the great fire of 1666, when Hooke was commissioned to help survey the City of London so that property could be rebuilt. He was paid handsomely by the landowners, but continued to live penuriously, "as was evident by a large Iron Chest of Money found after his Death, which had been lock'd down with the Key in it, with a date of the Time, by which it appear'd to have been so shut up for above thirty Years."[12]

Hooke's pendulum clock escapement saw universal use, as did countless other inventions of his, including the universal joint, that have helped our world go round smoothly ever since. His mechanism for regulating pocket watches and chronometers, based on the oscillation of a delicately coiled spring, regulated industry, commerce, and navigation for the next three hundred years. Although Hooke did not invent the microscope, he greatly improved it, and with the publication of his *Micrographia* in 1665, he established the cellular structure of living organisms and otherwise defined the field. He dabbled as an architect, designing the buildings that housed the Royal Society, the College of Physicians, the British Museum, and the Hospital of St. Mary of Bethlehem, or Bedlam. He is remembered for Hooke's law of elasticity and forgotten as the discoverer of the optical interference patterns known as Newton's rings.

When new inventions were presented to the Royal Society, Hooke either claimed to have invented them earlier or showed how they could be improved. When Leibniz exhibited a calculating machine in January 1673, Hooke complained, "it seemed to me so complicated with wheels, pinions, cantrights, springs, screws, stops, and truckles, that I could not perceive it ever to be of any great use. . . . It could only be fit for great persons to purchase, and for great force to remove and manage, and for great wits to understand and comprehend." In contrast, Hooke announced that "I have an instrument now making, which will perform the same effects [and] will not have a tenth part of the number of parts, and not take up a twentieth part of the room."[13] The record shows that on 5 March 1673, "he produced his arithmetical engine, mentioned by him in the meeting of February 5, and showed the manner of its operation, which was applauded." But Hooke's invention "whereby in large numbers, for multiplication or division, one man may be able to do more than twenty by the common way of working arithmetic" remained as sparsely documented as his spring-powered model representing one of some thirty different envisioned species of flying machines. The arithmetic engine was listed among the artificial rarities in the collection of Gresham College in 1681, and thereafter disappeared.

Hooke neglected most opportunities to reap reward. "Whether this mistake resulted from the multiplicity of his Business which did not allow him a sufficient time," wrote Richard Waller, "or from the fertility of his Invention which hurried him on, neglecting the former Discoveries . . . tho' there wanted some small matter to render their use more practicable and general, I know not."[14] If anyone could be said to have thought of everything, it was Hooke. He developed a philosophical algebra by which to grasp multiple avenues of thought at a single time, making it inevitable, even without unscrupulous behavior on the part of some of his colleagues, that competitors would appear to be taking credit for his ideas. His contributions included a theory of gravitation and celestial mechanics, of which, said Aubrey, "Mr. Newton haz made a demonstration, not at all owning he receiv'd the first Intimation of it from Mr. Hooke." Eventually, Hooke grew bitter over both real and perceived expropriations of his work and revealed many of his later inventions only in the form of cryptic anagrams, carrying the details with him to his grave. "I wish he had writt plainer, and afforded a little more paper," was Aubrey's chief complaint.[15]

Hooke was acquainted with the elder Thomas Hobbes, but "found him to lard and seal every asseveration with a round oath, to under-value all other men's opinions and judgments, to defend to the utmost what he asserted though never so absurd."[16] Their destinies were intertwined: it was Hooke's concept of long-distance communication that would bring Hobbes's Leviathan to life. Although Hooke did not go as far as Hobbes in assigning a material existence to the soul, he speculated more precisely on the physical operation of the mind.

The mystery, to Hooke, was not that we are able to perceive, remember, and generate new concepts from one moment to the next, but how the mind keeps track of temporal sequence while preserving random access to its store of memories and ideas. Hooke's solution—like the mechanism he developed for the regulation of chrono-meters—took the form of a coiled spring: "There is as it were a continued Chain of Ideas coyled up in the Repository of the Brain, the first end of which is farthest removed from the Center or Seat of the Soul where the Ideas are formed, which is always the Moment present when considered: And therefore according as there are a greater number of [layers of] these Ideas between the present Sensation or Thought in the Center, and any other, the more is the Soul apprehen-sive of the Time interposed."[17]

To estimate the storage capacity of the human brain, Hooke calculated the number of thoughts that could be registered per second, hour, day, year, and lifetime—"to take a round sum but 21 hundred Millions." He reduced to 100 million the number that the average person might remember, who "consequently must have as many distinct Ideas." Hooke then drew on his firsthand observations of microorganisms to argue that this many ideas might easily fit inside the brain: "I see no Reason why all these may not actually be contained within the Sphere of Activity of the Soul.... For if we consider in how small a bulk of Body there may be as many distinct living creatures as are here supposed Ideas, and every of these Creatures perfectly formed and endued with all its Vegetative and Animal Functions, and with sufficient room also left for it to move it self to and fro among and between all the rest . . . we shall not need to fear any Impossibility to find out room in the Brain."[18]

As early as 17 February 1664 the Royal Society urged "that Mr. Hooke set down in writing and produce to the Council his whole apparatus and management for speedy intelligence,"[19] but nothing was forthcoming until 29 February 1672, when "he proposed a way for a very speedy conveyance of intelligence from place to place by the sight assisted with telescopes, to be employed on high places, by the correspondents using a secret character.... The paper of this proposition, and the particulars of the manner of practising it, were read, but not left by Mr. Hooke to be registered, but taken away by him."[20] The council ordered "that some experiment should be made of this proposition at the next meeting," and on 7 March a test was conducted across the Thames. "The contrivance was applauded as very ingenious ... [but] the President objected, that the use of it would be often hindered by hazy weather."[21]

In a disclosure that was finally delivered and recorded in 1684, Hooke prescribed an alphabet of twenty-four symbols, constructed of thin wood and rigged via pulleys and control lines so as to be exposed as required from behind an elevated wooden screen. "By these Contrivances, the Characters may be shifted almost as fast, as the same may be written; so that a great Quantity of Intelligence may be, in a very short Time, communicated."[22] For nighttime use, Hooke proposed a 2×5 array of lanterns "disposed in a certain Order, which may be veiled, or discovered, according to the Method of the Character agreed on; by which, all Sorts of Letters may be discovered clearly, and without Ambiguity,"[23] foreshadowing the shutter telegraphs that would be instituted by the British Admiralty in another hundred years. Finally, confirming the intimate association between telecommunications and cryptography, Hooke noted that by "cruptography" (as he spelled it) the arbitrary mapping between symbols and letters permits "the whole alphabet [to] be varied 10,000 ways; so that none but the two extreme correspondents shall be able to discover the information conveyed."[24]

Hooke specified single-character control codes to be displayed above the message area during transmission, providing eleven examples of these out-of-band signals, of which Holzmann and Pehrson noted that "at least eight are control codes that can be found in most modern data communication protocols, and some of these (i.e., the rate control codes) only in the most recent designs."[25] Hooke confidently predicted that "things may be made so convenient, that the same Character may be seen at *Paris*, within a Minute after it hath been exposed at *London*, and the like in Proportion for greater Distances; and that the Characters may be exposed so quick after one another, that a Composer shall not much exceed the Exposer in Swiftness."[26]

By the end of the next century, optical telegraph networks spanned most of Europe, led by a system constructed by Claude and Abraham Chappe in France. In 1801, with designs on England, Napoléon commissioned an optical telegraph able to span the English Channel; when tested over equivalent distances, it worked. Claude Chappe (1763–1805) had attempted to construct an electric telegraph in 1790, but soon abandoned electricity in favor of optical signals relayed by mechanical display. The French Revolution was in full swing. The new government was open to new ideas, but Chappe's prototype installation was destroyed twice by revolutionary mobs who suspected it was a device for communicating with the imprisoned Louis XVI. Chappe's network, inaugurated by a 130-mile line between Paris and Lille in 1794, reached a total length of approximately 3,000 miles, staffed by some three thousand operators at 556 stations, in 1852. Stations were about 6 miles apart. Signals could be relayed in a few seconds, but the lines ran much slower in practice, with actual throughput about two signals per minute or less. Transit over the 475 miles and 120 stations from Paris to Toulon (on the Mediterranean) took "10 or 12 minutes" if the weather was clear. Messages were encrypted against interception or adulteration en route.

Chappe's coding improved on the one-to-one mapping between transmitted symbols and written alphabet proposed by Hooke. Two independently rotating indicators (each capable of seven distinguishable positions) were mounted at the ends of a central regulator that alternated between two positions. There were thus ninety-eight ($7 \times 7 \times 2$) recognizably different regulator/indicator positions, of which six were reserved for special indications, leaving ninety-two signals for conveying the message code. Taking a page from Polybius, the Chappes instituted a ninety-two-page codebook with ninety-two words or phrases on each page. The first page encoded the alphabet, numerals, and most frequently transmitted syllables, which could be transmitted as a single signal followed by folding in the indicators "closed." The following ninety-one pages, each listing ninety-two words and phrases, were composite signals, transmitted as two consecutive signals designating page number and line.

This system allowed 8,464 different meanings to be encoded as signal pairs. In 1799 the code space was increased to a total of 25,392 entries with the addition of two auxiliary books. A large number of meanings were represented by the combination of comparatively few symbols, using some of the same principles of data compression prevalent today. As Alan Turing would later demonstrate by the design of his theoretical Turing machine, any arbitrarily complicated symbol, or sequence of symbols, could be captured as a state of mind, while the representation of any state of mind could be transmitted by running the process the other way. The codebooks developed by the Chappes were a concrete example of a computable function relating a given scanned symbol or short sequence of symbols to an equivalent state of mind.

Chappe's system was imitated, with modifications, throughout the developed world. In Russia, 1,320 people were employed just to operate the main line from St. Petersburg to the Prussian frontier. In England, the 200-mile line between London and Plymouth was able to convey a timing signal and return an acknowledgment in under three minutes when the weather was clear. Fragments of the optical telegraph networks survived, especially in island-studded Scandinavia, for many years after electric telegraphy was introduced. According to Holzmann and Pehrson, Claude Chappe and his Swedish counterpart, Abraham Edelcrantz, were the "true pioneers of data networking" and managed "to solve many subtle problems to enable operators to transfer messages smoothly over long chains of stations ... ideas [that] have been rediscovered only recently by the designers of modern digital protocols."[27] In a fitting reprise of the original message transmitted to Mycenae from Troy, one of the last messages distributed in part by optical telegraph reported the fall of Sevastopol during the Crimean War in 1855. Electric telegraphs, however, were already in use by the British in the final stage of the siege, and on the Russian side electrical connections reaching back to St. Petersburg were already complete as well.

The development of the electric telegraph during the first half of the nineteenth century represented the culmination of principles incubated over many years. As early as July 1729, Stephen Gray of London transmitted an electric charge a distance of 765 feet. In July 1747, following the invention of the Leyden jar, or charge-storing capacitor, in 1745, William Watson succeeded in communicating an electric charge by iron wire across the Thames at Westminster Bridge; in 1748, Benjamin Franklin did the same across the Schuylkill River; on 5 August 1748, Watson and Henry Cavendish extended the distance to 12,276 feet. The Abbé Jean Antoine Nollet (1700–1770) succeeded in electrifying 180 of the king's palace guards, who "felt the shock at the same time" at Paris in 1746. Later, using lengths of iron wire, he formed a 900-foot chain of Carthusian monks and reported that "the whole company at the same instant of time gave a sudden spring."[28]

In February 1753, a Scottish inventor identified only as C.M. (believed to have been either one Charles Marshall, of Paisley, or Charles Morrison, of Renfrew) described "an expeditious method for conveying intelligence" by means of twenty-four parallel wire conductors supported by glass insulators about "every twenty yards." Observing that "electric power may be propagated along a small wire, from one place to another, without being sensibly abated by the length of its progress," C.M. detailed the construction and operation of an electric telegraph, though there is no evidence that the machine was ever built. "Let the wires be fixed in a solid piece of glass, at six inches from the end; and let that part of them which reaches from the glass to the machine have sufficient spring and stiffness to recover its situation after having been brought in contact with the barrel." By depressing individual conductors in sequence, any alphabetic message could be conveyed, or, by reversing the process, on an agreed signal, a reply could be received. "Close by the supporting glass, let a ball be suspended from every wire; and about a sixth or an eighth of a inch below the balls, place the letter of the alphabet, marked on bits of paper, or any other substance that may be light enough to rise to the electrified ball; and at the same time let it be so contrived, that each of them may reassume its proper place when dropt." Less practical was the suggestion to employ electrically actuated bells, by means of which the two correspondents "may come to understand the language of the chimes in whole words."[29]

Telegraphs surfaced like weeds. In his *History of Electric Telegraphy to the Year 1837,* John J. Fahie reviewed the work of some forty-seven inventors whose telegraphs succumbed to technical obstacles, lack of support, or a combination of both. In the 1790s, Don Francisco Salvá of Barcelona (1751–1828) enjoyed the support of the king and was rumored to have constructed a single-wire telegraph line over the twenty-six miles between Aranjuez and Madrid. Salvá experimented both with electrostatic signals and with the transmission of faint pulses of direct current, indicated by the convulsion of frog legs as much as 310 meters apart. Working before news of Volta's battery was received, "Salvá employed, as his motive power," reported Fahie, "the electricity produced by a great number of frogs."[30] He delivered a paper entitled "Galvanism and its application to Telegraphy" on 14 May 1800 to the Academy of Sciences in Barcelona, followed by a second treatise in 1804 showing how the frogs could be replaced as both transmitters and receivers of electric signals by electrochemical cells. In England, Francis Ronalds of Hammersmith, London, working at his own expense, transmitted electrostatic signals over eight miles of wire in 1816. Envisioning "electrical conversazione offices, communicating with each other all over the kingdom,"[31] but foreseeing the vulnerability of overhead lines, he proposed a network of buried cables, testing his apparatus over 525 feet of insulated conductor buried 4 feet underground. He then wrote to the first lord of the Admiralty, "soliciting his lordship's attention to a mode of conveying telegraphic intelligence with great rapidity, accuracy, and certainty, in all states of the atmosphere, either at night or in the day, and at small expense."[32] The official response was that "telegraphs of any kind are now wholly unnecessary, no other than the one now in use will be adopted.[33] A section of Ronalds's cable was excavated intact in 1862. Success depends on timing—not necessarily on being first.

In March 1800, Alessandro Volta announced his invention of the voltaic pile to the Royal Society, and electric batteries were duplicated around the world. In 1819, Hans Christian Oersted outlined the essential principles of electromagnetism, brought into coherent mathematical form by André-Marie Ampère, who delivered his first paper on electrodynamics, as he termed it, on 18 September 1820, after only seven days of work. In October 1820, some fifteen years before he first put the word *cybernétique* into print, Ampère followed a suggestion of his colleague Pierre Laplace in proposing an electric telegraph. "One could, by means of as many conducting wires and magnetic needles as there are letters, and by placing each letter on a different needle, establish by the aid of a voltaic pile . . . a genuine telegraph, writing all the details one might wish to transmit, across whatever obstacles," noted Ampère. "By connecting to the voltaic pile a keyboard whose keys bear the same letters and establish contact by their depression, this method of correspondence could be performed with great facility, in no more time than necessary to touch each letter at one end, and read it at the other."[34]

Commercially successful telegraphy slowly evolved. Electrochemical indicators that displayed signals by releasing bubbles of gas were invented, as was a system that inscribed electrostatic signals as marks on a moving strip of litmus paper, demonstrated by Harrison Gray Dyar over 8 miles of wire encircling a Long Island racetrack in 1827. Construction of a line between New York and Philadelphia was terminated when charges of bank conspiracy brought the project's financing to a halt. A five-channel needle telegraph was developed by the Russian baron Paul L. Schilling in 1823 and exhibited as a working model to Czar Alexander I in 1825. Carl Friedrich Gauss and Wilhelm E. Weber constructed a 1.5-mile galvanometer telegraph in Göttingen in 1833, used daily to communicate between the physics department and the observatory until 1838. In the United States, there were the proof-of-principle experiments of Joseph Henry, appointed first secretary of the Smithsonian Institution in 1846, who rang a bell at a distance of 1,000 feet in 1830, followed by a 1-mile relay-actuated communication link between his house in Princeton, New Jersey, and Princeton College, where he worked.

Returning in 1832 from Europe, where electrical developments were in the air, Samuel Morse entertained visions of electromagnetic telegraphy, making his first notes toward the binary dot-dash representation of numbers and letters later known as the Morse code. A professor of art history, not an electrical engineer, Morse relied on the assistance of Joseph Henry and Alfred Vail to develop a system that worked. Vail became rich, while Henry, who asked for nothing, was personally attacked when the circumstances surrounding his contributions threatened to undermine the patents held by Morse. The Morse telegraph incorporated relays for pulse regeneration, closely following Henry's lead. Telegraph relays, manufactured in large quantities and adapted for switching purposes, would spark the proliferation of binary logic gates that led to the age of digital machines.

Morse established the first long-distance line, between Baltimore and Washington, which opened on 24 May 1844 with a message selected by Miss Annie Ellsworth, daughter of the commissioner of patents—the last phrase of the twenty-third verse of the twenty-third chapter of the Book of Numbers: "What hath God wrought!" Success attracted both converts and competitors, and by 1851, the year the progenitor of Western Union was established, there were over fifty telegraph companies in the United States. That same year the first telegraph cable linked England and France; by 1852 there were some twenty-three thousand miles of telegraph lines in existence, enough to encircle the world. In 1861 the first line spanned the North American continent, and in 1866, after many failures, a durable connection linked England to the United States. India was reached in 1870, Australia in 1871, and in 1874, Brazil.

The obstacles shifted from establishing the physical connections constituting each leg of a telegraph circuit to switching, regenerating, and encoding and decoding the messages at either end. Telegraph signals were digital signals, whether conveyed by the on–off state of a fire beacon, the twenty-four-symbol alphabet of Robert Hooke, the ninety-eight-state signal of the Chappes, a series of positive–negative voltages, or the dot-dash sequences of Morse code. To process these signals requires discrete-state machines, whether the machine is a human operator looking through a telescope and referring to page and line numbers in a book of code or one of the punched-tape teleprinters that soon came to dominate telegraphy throughout the world.

Telegraph engineers were the first to give substance to what had been shown by Leibniz two centuries earlier and would be formalized by Alan Turing in the century ahead: all symbols, all information, all meaning, and all intelligence that can be described in words or numbers can be encoded (and thus transmitted) as binary sequences of finite length. It makes no difference what form the symbols take; it is the number of choices between alternatives that counts. It takes five binary alternatives ($2^5 = 32$) to encode the alphabet, which is why early needle telegraphs used five separate two-state indicators and why teletypewriter tape is five holes wide. (Polybius had specified two separate five-state indicators, a less efficient coding that makes sense if you are keeping a naked eye out for torches in the dark.)

Every message had to be encoded, decoded, stored, reencoded, and retransmitted many times as it made its way from one node to the next. In 1858, Charles Wheatstone introduced perforated paper tape as a means of automatic signal transmission; receiving perforators, reperforators, and perforated-tape-driven printers soon followed. In the 1870s, Jean Maurice Émile Baudot introduced both time-division multiplexing (interweaving several code sequences over a single circuit) and the 5-bit alphanumeric code that bears his name (Wilkins was long forgotten by this time). Although beginning and ending its journey as alphabetic text, the message was represented as either sequences of pulses over a wire or sequences of punches in a strip of paper through the many stages along the way.

The telegraph system soon evolved store-and-forward procedures at its nodes—the ancestor of the packet-switching protocols used in computer networks today. An incoming telegram arrived at the switching node as a sequence of electrical signals, converted to a series of holes in a strip of paper tape, identified by its origin, its priority, and its destination address. The station operator considered this information along with the state of the outgoing lines in making a decision as to when and via what routing to relay the message, retranslated back into electrical pulses by a machine that sensed the pattern on the tape. When the message was acknowledged by the next station, the tape could be discarded, and the transitory state of mind that it represented was erased.

High-speed automatic telegraph instruments were the ancestors of modern computers and gave the electromagnetic data-processing industry its start. "Computing machines," explained John von Neumann in 1949, "can be thought of as machines which are fed, and emit, some medium like punched tape."[35] This definition works both ways. Like the molecules that convey hereditary information between living cells, telegraph equipment performed the function of recording, storing, and transferring sequences of code. Most early digital computers—from the Colossus to the IAS machine—used paper-tape teletype equipment for input and output between the computer and the outside world, augmented by the ubiquitous punched-card equipment from Hollerith and IBM. It was only natural that the first computers incorporated high-speed telegraphic equipment, and it is no accident that the genesis of the Colossus within the British Telecommunications Research Establishment was mirrored in the United States by early steps taken toward computers by Claude Shannon and others within Bell Laboratories and RCA. Only later did the communications industry and the computer industry become temporarily estranged.

The solitary computers of the early 1950s exchanged code sequences by means of mutually intelligible storage media and, before the end of the decade, by connecting directly or, in language that has now been extended to human beings, on-line. But no matter what the medium of exchange, the code itself and the protocols that regulate its flow remain directly descended from the first strings of telegraphic bits. Evolution of computer-to-computer communication, like previous advances in telecommunications, was closely related to defense. In the early 1950s, a computer project known as Whirlwind evolved to support an integrated air-defense system developed for the U.S. Air Force by the Lincoln Laboratory at MIT. Whirlwind led directly to the SAGE (Semi-Automatic Ground Environment) air-defense system, a real-time interactive data-processing system, which led in turn to the development of time-sharing computer systems and, eventually, to computer networks as we know them today. John von Neumann's legacy to computer networking was to be found not only in the architecture of individual computers, but in the proliferation of weapons against which networks of computers offered the best hope of defense.

In 1955, less than three years after the explosion of the first hydrogen bomb, von Neumann was able to announce that "we can pack in one airplane more firepower than the combined fleets of all the combatants during World War II."[36] Von Neumann was also thinking of long-range missiles, which, as chairman of the Strategic Missiles Evaluation Committee established in 1953, he referred to as "nuclear weapons in their expected most vicious form."[37] Lightweight, high-yield thermonuclear weapons would soon be available that could be lofted by rocket into space and guided to earth many thousands of miles away. After the Soviets exploded their first hydrogen bomb in 1953, the push to develop an intercontinental ballistic missile, or ICBM, became a race.

Coaching the U.S. team was a nonprofit organization known as RAND. Incorporated in 1948 "to further and promote scientific, educational, and charitable purposes, all for the public welfare and security of the United States of America,"[38] RAND (Research And Development) was the successor to U.S. Air Force Project RAND, established by a contract of 2 March 1946 specifying that "the Contractor will perform a program of study and research on the broad subject of intercontinental warfare, other than surface, with the object of recommending to the Army Air Forces preferred techniques and instrumentalities for this purpose."[39] Headquartered in Santa Monica, California, RAND was constituted as a refuge for the free-thinking academic approach to military problems that had thrived during the war but risked being extinguished by the peace.

Although operating on principles similar to those of the Institute for Advanced Study, RAND went about the business of creative thinking in reverse. The Institute, established for the pursuit of pure research, quietly facilitated military work. RAND was openly targeted at military objectives, while quietly facilitating pure science and advancing scientific careers. Cross-fertilization between pure and applied mathematics flourished at RAND as nowhere else. In 1955, RAND published *A Million Random Digits with 100,000 Normal Deviates,* whose introduction notes that "because of the very nature of the tables, it did not seem necessary to proofread every page of the final manuscript in order to catch random errors."[40] In the 1950s there was a serious shortage of random numbers, and RAND's random numbers were widely distributed and used for solving problems in many fields. RAND researchers were responsible for writing and defending their own reports. The resulting publications, aimed at air force generals, not academic colleagues, were distinguished by a clarity, economy, and self-contained documentation rarely seen in academic work.

RAND's first published study was the 324-page *Preliminary Design of an Experimental Earth-Circling Spaceship,* issued on 2 May 1946. The report advised the government that "the achievement of a satellite craft by the United States would inflame the imagination of mankind, and would probably produce repercussions in the world comparable to the explosion of the atomic bomb. . . . Whose imagination is not fired by the possibility of voyaging out beyond the limits of our earth, traveling to the Moon, to Venus and Mars?"[41] When the Soviets launched the first *Sputnik,* on 4 October 1957, the American public was taken by surprise, but it was no surprise to RAND, whose analysts had predicted the appearance of a Soviet satellite on 17 September 1957, the centenary of the birth of Konstantin Tsiolkovsky, the great Russian rocket pioneer. In November 1957, after *Sputnik II* had circled the earth with a payload of 1,120 pounds (including the dog Laika), there was no longer any doubt that the Soviets were thinking about launching not only dogs, but bombs as well. U.S. spending on the Atlas ICBM program jumped from $3 million in 1953 to $161 million in 1955 and $1,300 million in 1957. RAND studies led the way. "By 1953 RAND's knowledge of the missile and weapon fields indicated that nuclear warheads could be carried by rockets, and could produce a wide enough zone of destruction to more than compensate for their aiming errors," reported a RAND history in 1963.[42]

With weapons multiplying on both sides, RAND sought to identify stable nuclear strategies, hosting a renaissance of mathematical game theory beginning where von Neumann and Morgenstern had left off. The titles of some published reports convey the general drift: *The noisy duel, one bullet each, arbitrary nonmonotone accuracy* (D. H. Blackwell, March 1949); *A generalization of the silent duel, two opponents, one bullet each, arbitrary accuracy* (M. A. Girshick, August 1949); *A loud duel with equal accuracy where each duelist has only a probability of possessing a bullet* (M. A. Girshick and D. H. Blackwell, August 1949); *The silent duel, one bullet versus two, equal accuracy* (L. S. Shapley, September 1950); *Noisy duel, one bullet each, with simultaneous fire and unequal worths* (I. L. Glicksberg, October 1950).

The Distant Early Warning (DEW Line) system constructed in 1953 offered a two- or three-hour warning of Soviet bomber attack. By the time the Ballistic Missile Early Warning system came on-line in 1960, the United States was reduced to fifteen or thirty minutes' warning of an incoming missile attack. Nuclear stability depended on mutual deterrence, either by threatening to launch missiles at the first sign of enemy attack or by preserving the ability to retaliate after a strike. Launch-on-warning, serviceable as a bluff, would be suicidal in practice, since sooner or later an erroneous warning would arise. So the best way to prevent a nuclear nightmare appeared to be to construct a retaliatory system designed to survive attack. To hide, disperse, or harden missiles was comparatively easy; the difficulty was how to construct a robust system of control.

"Throughout history successful generals make their plans based upon enemy capabilities and not intent," recalled Paul Baran, an electrical engineer who arrived at RAND at age thirty-three in 1959. "This period was the height of the cold war. Both the U.S. and USSR were building hair-trigger nuclear ballistic missile systems. The early missile-control systems were not physically robust. Thus, there was a dangerous temptation for either party to misunderstand the actions of the other and fire first. . . . If we could wait until after attack rather than having to respond too quickly under pressure then the world would be a more stable place." Although use of the word *surrender* was forbidden by act of Congress, RAND officials recognized that "a survivable communications network is needed to stop, as well as to help avoid, a war."[43]

A 1960 RAND study, *Cost of a Hardened, Nationwide Buried Cable Network,* estimated the cost of providing two hundred facilities with communication links hardened against 100-psi blast pressure at $2.4 billion, with protection to 1,000 psi available for about $1 billion additional cost.[44] Baran was assigned the job of analyzing whether it was possible to do as well or better for less. Retaliatory ability could be met by an extremely low bandwidth channel, or "minimal essential communications"—official terminology for the president's (or his successor's) ability to issue the order "Fire your missiles" or "Stop."

Baran began with Frank Collbohm's suggestion that by installing a minimal amount of digital logic throughout the existing nationwide network of AM radio stations, a decentralized and highly redundant communication channel could be engaged in the event of an attack. The strategy was to flood the network with a given message. No routing was required other than a procedure that ceased transmission of a specific message once copies of it started bouncing back. Analysis showed that even with widespread destruction of network nodes, with metropolitan broadcast facilities disappearing first, the overlapping nature of the system would allow messages to diffuse rapidly throughout the surviving stations on the net (an unstated conclusion being that American country music could survive even a worst-case Soviet attack). Baran took this proposal to the armed forces, which held out for more bandwidth, believing that real-time voice communication was required to fight a war. "Okay, back to the drawing board," said Baran, "but this time I'm going to give them so much damn communication capacity they won't know what in hell to do with it all."[45] And he did.

Baran's first job had been with the Eckert–Mauchly Computer Company in 1949. The weaknesses of vacuum-tube–delay-line computers at that time suggested a tenuous future for such unwieldy and temperamental machines. In ten years everything had changed.

Witnessing this revolution encouraged Baran to question other assumptions as well. The creative approach to digital computing that had flourished in the early 1950s at the Institute for Advanced Study continued to flourish in the early 1960s at RAND. Baran's thesis advisor at UCLA was Gerald Estrin, who had been instrumental in disseminating the computer project at the IAS. JOHNNIAC, the RAND version of the IAS machine, incorporated several improvements, including a working Selectron memory, and was completed under the direction of Willis Ware, also an alumnus of the IAS. RAND had steadily acquired the latest machines from IBM. Working under the auspices of the computer department, not the communications department, Baran's analysis of communication problems was able to develop unencumbered by preconceived ideas. "Computers and communications," he remarked, "were at that time two totally different fields."[46]

Baran invented a new species of communications network, starting with clear-cut objectives and little else. "An *ideal* electrical communications system can be defined as one that permits any person or machine to reliably and instantaneously communicate with any combination of other people or machines, anywhere, anytime, and at zero cost," he wrote. "It should effectively allow the illusion that those in communication with one another are all within the same sound-proofed room—and that the door is locked."[47] He threw out all existing assumptions. "In most communication applications, silence is the *usual* message," he later explained.[48]

By digitizing all communications and multiplexing across the entire network rather than over one channel at a time, Baran knew he could reduce most of the waste. Taking a cue from the store-and-forward torn-tape telegraph networks, he proposed relaying digital messages from node to node across the net, but with high-speed computers instead of telegraph equipment providing switching and storage at the nodes. He examined existing military communications switches and asked: "Why are these things so big? Why do these switching systems require rooms and rooms of stuff?" He concluded that there was far too much recording and storage of messages at the switching nodes, for no apparent reason other than a tradition "to be able to prove that lost traffic was someone else's fault."[49] Computers were already operating at multimegacycle rates, and Baran knew that the telecommunications infrastructure would eventually be forced to catch up. He proposed a system operating at up to 1.5 million bits per second over low-power, line-of-sight microwave links—ample bandwidth for real-time transmission of digitally encrypted voice. This proposal secured the undivided enthusiasm of the military commanders, if guaranteeing stiff resistance from the managers of AT&T, which handled all military voice communications within the United States. They were not about to admit that their system was physically vulnerable, inefficient, or insecure.

In May 1960, RAND released the first official memorandum that provided an outline of Baran's design. "If war does not mean the end of the earth in a black and white manner, then it follows that we should do those things that make the shade of grey as light as possible," wrote Baran in his introduction to *Reliable Digital Communications Systems Utilizing Unreliable Network Repeater Nodes*. "We are just beginning to design and lay out designs for the *digital* data transmission systems of the future ... systems where computers speak to each other. ... As there does not seem to be any fundamental technical problem that prohibits the operation of digital communication links at the clock rate of digital computers, the view is taken that it is only a matter of time before such design requirements become hardware ... where the intelligence required to switch signals to surviving links is at the link nodes and *not* at one or a few centralized switching centers."[50]

Baran then suggested how to evaluate the robustness of his design: "To better visualize the operation of the network, a hypothetical application is postulated: a congressional communications system where each congressman may vote from his home office. The success of such a network may be evaluated by examining the number of congressmen surviving an attack and comparing such number to the number of congressmen able to communicate with one another and vote via the communications network. Such an example is, of course, farfetched but not completely without utility."[51]

The more alternative connection paths there are between the nodes of a communications net, the more resistant it is to damage from within or without. But there is a combinatorial explosion working the other way: the more you increase the connectivity, the more intelligence and memory is required to route messages efficiently through the net. In a conventional circuit-switched communications network, such as the telephone system, a central switching authority establishes an unbroken connection for every communication, mediating possible conflicts with other connections being made at the same time. "Such a central control node offers a single, very attractive target in the thermonuclear era," warned Baran.[52] The stroke of genius at the heart of Baran's proposal was to distribute the requisite intelligence and redundancy not only among the individual switching nodes, but also among the messages themselves.

Baran credited the ancestry of this approach to Theseus, a mechanical mouse constructed by information theorist Claude Shannon in 1950. Guided by the intelligence of seventy-two electromagnetic relays, Theseus was able to find its way around a 5 × 5 maze, in Warren McCulloch's words, "like a man who knows the town, so he can go from any place to any other place, but doesn't always remember how he went."[53] To adapt to changes in the layout of the maze and the location of the goal, Theseus had to be able not only to remember but to forget. Baran saw that the problem of routing a mouse through a maze was equivalent to the problem of routing messages through a communications net. "In a very short period of time—within the past decade, the research effort devoted to these ends has developed from analyses of how a mechanical mouse might find his way out of a maze, to suggestions of the design of an all-electronic world-wide communications system," he wrote in 1964.[54]

Baran christened his technique "adaptive message block switching," abbreviated to "packet switching" in 1966 by Donald Davies, working independently at the U.K. National Physical Laboratory. The first order of business was to take all forms of communicable information—text, data, graphics, voice—and break it up into short strings of bits of uniform length. To the network, all forms of communication would look the same. Error-free transmission of complex messages would be facilitated, for the same reason that the reproduction of complex organisms can best be achieved in a noisy environment through the collective reproduction of large numbers of smaller component parts. Every bit of data that passes through a network acquires a certain probability of error, and the cumulative probability of one of these errors affecting any given message grows exponentially with the message length. Short segments of code are far more likely to arrive unscathed. Error-detecting processes are most economically applied to short strings of code, checking each message segment for errors and only retransmitting those individual segments that fail. This is analogous to proofreading a manuscript one page at a time and retyping only those pages containing mistakes. Baran proposed using cyclical redundancy checking, or CRC, a method widely utilized today.

Each 1,024-bit message block was flagged to distinguish it from its neighbors and provided with fields containing its "to" and "from" address, as well as information needed to reconstruct the original message sequence at the other end. The message block also included a handover tag, which was incremented every time the code segment passed through a node. Each individual code packet knew where it was going, where it had come from, which message it belonged to, and how many steps it had taken along the way. This information was shared with the host computer every time a packet passed through a node. "Packets are best viewed as wholly contained information entities designed to match the characteristics of the network switches," noted Baran.[55]

At the nodes, simple procedures (dubbed the "hot potato heuristic routing doctrine" by Baran) kept packets moving in the right direction and ensured that the network would adjust itself to any congestion or damage that arose. By looking at the "from" address and the value of the handover tag, the station kept track of which links were forwarding messages most efficiently from any given address and used that information to direct outgoing messages, operating on the assumption that the best link coming in was likely to be the best link going out. "Each node will attempt to get rid of its messages by choosing alternate routes if its preferred route is busy or destroyed. Each message is regarded as a 'hot potato,' and rather than hold the 'hot potato,' the node tosses the message to its neighbor, who will now try to get rid of the message."[56]

Individual switching nodes learned from experience where each station was and quickly updated this knowledge if a given station was damaged or moved. Baran found it possible to "suggest the appearance of an adaptive system" by implementing a "self-learning policy at each node so that overall traffic is effectively routed in a changing environment—without need for a central and possibly vulnerable control."[57] A demonstration system, simulated as a forty-nine-node array on RAND's IBM 7090 computer, proved to be surprisingly robust against both random failure and deliberate attack. Starting from a "worst-case starting condition where no station knew the location of any other station," it was found that "within ½ second of simulated real world time, the network had learned the locations of all connected stations and was routing traffic in an efficient manner. The mean measured path length compared very favorably to the absolute shortest possible path."[58]

The complete study was published in a series of eleven reports released in August 1964. An additional two volumes on cryptographic issues remained classified, even though, as Baran explained in the ninth volume, devoted to security and secrecy, "if one cannot safely describe a proposed system in the unclassified literature, then, *by definition*, it is not sufficiently secure to be used with confidence."[59] Baran later emphasized that "we chose not to classify this work and also chose not to patent the work. We felt that it properly belonged in the public domain. Not only would the US be safer with a survivable command and control system, the US would be even safer if the USSR also had a survivable command and control system as well!"[60]

In late 1962 it was estimated that each switching node would occupy seventy-two cubic feet, consume fifteen hundred watts of power, weigh twenty-four hundred pounds, and incorporate 4,096 32-bit words of memory. By 1964 the estimate was down to units occupying one-third of a cubic foot, with twice the memory, no air-conditioning, and costing fifty thousand dollars less. Unlike most military budgets, this one was going down. By the time the final reports were released, most skeptics had been persuaded of the project's advantages—"the only major outpost of frontal opposition was AT&T."[61] In 1965, RAND gave the air force its final recommendation to implement the project in a preliminary configuration comprising two hundred multiplexing stations distributed across the continental United States, each capable of serving up to 866 simultaneous subscribers with data and 128 with voice. There were to be four hundred switching nodes, each supporting up to eight full-duplex "minicost" microwave links. By 1966 the air force had received a favorable evaluation from an independent review. Only then was it determined that jurisdiction over the project would have to be assigned not to the air force and its independent contractors but to the Defense Communications Agency, which Baran and some influential colleagues believed was ill prepared to construct a nationwide network based on digital principles at odds with the communications establishment of the time. Baran was forced to advise against the implementation of his own project, lest "detractors would have proof that it couldn't be done . . . a hard decision, but I think it was the right one."[62]

Baran's packet-switched data network did eventually materialize— not out of whole cloth, as envisioned in 1960, but by the gradual association of many different levels of digital communication systems that ultimately converged, more or less closely, on Baran's original design. "The process of technological development is like building a cathedral," said Baran. "Over the course of several hundred years: new people come along and each lays down a block on top of the old foundations, each saying, 'I built a cathedral.' Next month another block is placed atop the previous one. Then comes along an historian who asks, 'Well, who built the cathedral?' But the reality is that each contribution has to follow onto previous work. Everything is tied to everything else."[63]

THEORY OF GAMES AND ECONOMIC BEHAVIOR

The game that nature seems to be playing is difficult to formulate. When different species compete, one knows how to define a loss: when one species dies out altogether, it loses, obviously. The defining win, however, is much more difficult because many coexist and will presumably for an infinite time; and yet the humans in some sense consider themselves far ahead of the chicken, which will also be allowed to go on to infinity.

—STANISLAW ULAM[1]

"Unifications of fields which were formerly divided and far apart," counseled John von Neumann in 1944, "are rare and happen only after each field has been thoroughly explored."[2] So went his introduction (with Oskar Morgenstern) to *Theory of Games and Economic Behavior,* a mathematical vision whose brilliance was eclipsed only by the developments that his work in atomic weapons and digital computers was about to bring to light.

An interest in economics ran deeply through von Neumann's life. At the time of the Institute for Advanced Study's electronic computer project, von Neumann maintained an incongruous appearance, wearing a three-piece suit among casually dressed logicians and electrical engineers. This costume was a memento of his background as the son of an investment banker and the omen of a future in which the world of money and the world of logic, thanks to computers, would meet on equal terms. In his *Theory of Games and Economic Behavior,* von Neumann laid the foundations for a unified view of information theory, economics, evolution, and intelligence, whose implications continue to emerge.

Among von Neumann's predecessors was André-Marie Ampère, who published *Considérations sur la théorie mathématique du jeu* (On the mathematical theory of games) at the age of twenty-seven in 1802. Ampère began his study by crediting Georges Louis Buffon ("an author in whom even errors bear the imprint of genius") as the forefather of mathematical game theory, citing his (1777) *Essai d' Arithmétique Morale.* Buffon (1707–1788) was a celebrated naturalist whose evolutionary theories preceded both Charles and Erasmus Darwin, advancing ideas that were risky at the time. "Buffon managed, albeit in a somewhat scattered fashion," wrote Loren Eiseley, "at least to mention every significant ingredient which was to be incorporated into Darwin's great synthesis of 1859."[3] The elder Buffon and the young Ampère shared in the tragedy that swept postrevolutionary France: Buffon's son and Ampère's father both died under the guillotine, equally innocent of any crime.

Ampère analyzed the effects of probability rather than strategy, ignoring more deliberate collusion among the players of a game. Having suffered the first of a series of misfortunes that would follow him through life, Ampère saw games of chance as "certain ruin" to those who played indefinitely or indiscriminately against multiple opponents, "who must then be considered as a single opponent whose fortune is infinite."[4] He observed that a zero-sum game (where one player's loss equals the other players' gain) will always favor the wealthier player, who has the advantage of being able to stay longer in the game.

Von Neumann's initial contribution to the theory of games, extending the work of Émile Borel, was published in 1928. Where Ampère saw chance as holding the upper hand, von Neumann sought to make the best of fate by determining the optimum strategy for any game. The results of his collaboration with Princeton economist Oskar Morgenstern were completed in the midst of wartime and published in 1944. "The main achievement of the [von Neumann–Morgenstern] book lies, more than in its concrete results, in its having introduced into economics the tools of modern logic and in using them with an astounding power of generalization," wrote Jacob Marschak in the *Journal of Political Economy* in 1946.[5] Von Neumann's central insight was his proof of the "minimax" theorem on the existence of good strategies, demonstrating for a wide class of games that a determinable strategy exists that minimizes the expected loss to a player when the opponent tries to maximize the loss by playing as well as possible. This conclusion has profound but mathematically elusive consequences; many complexities of nature, not to mention of economics or politics, can be treated formally as games. A substantial section of the 625-page book is devoted to showing how seemingly intractable situations can be rendered solvable through the assumption of coalitions among the players, and how non-zero-sum games can be reduced to zero-sum games by including a fictitious, impartial player (sometimes called Nature) in the game.

Game theory was applied to fields ranging from nuclear deterrence to evolutionary biology. "The initial reaction of the economists to this work was one of great reserve, but the military scientists were quick to sense its possibilities in their field," wrote J. D. Williams in *The Compleat Strategyst,* a RAND Corporation best-seller that made game theory accessible through examples drawn from everyday life.[6] The economists gradually followed. When John Nash was awarded a Nobel Prize for the Nash equilibrium in 1994, he became the seventh Nobel laureate in economics whose work was influenced directly by von Neumann's ideas. Nash and von Neumann had collaborated at RAND. In 1954, Nash authored a short report on the future of digital computers, in which the von Neumann influence was especially pronounced. "The human brain is a highly parallel setup. It has to be," concluded Nash, predicting that optimal performance of digital computers would be achieved by coalitions of processors operating under decentralized parallel control.[7]

In 1945 the *Review of Economic Studies* published von Neumann's "Model of General Economic Equilibrium," a nine-page paper read to a Princeton mathematics seminar in 1932 and first published in German in 1937. With characteristic dexterity, von Neumann managed to elucidate the behavior of an economy where "goods are produced not only from 'natural factors of production,' but . . . from each other," thereby shedding light on processes whose interdependence otherwise appears impenetrably complex. Because equilibrium was shown to depend on growth, this became known as von Neumann's expanding economic model. The conclusions were universally debated by economists; mathematicians were universally impressed. Von Neumann derived his conclusions from the topology of convex sets, noting that "the connection with topology may be very surprising at first, but the author thinks that it is natural in problems of this kind."[8]

Von Neumann was laying the groundwork for a unified theory of information dynamics, applicable to free-market economies, self-reproducing organisms, neural networks, and, ultimately, the relations between mind and brain. The confluence of the theory of games with the theory of information and communication invites the construction of such a bridge. In his notes for a series of lectures that were preempted by his death, von Neumann drew a number of parallels—and emphasized a greater number of differences—between the computer and the brain. He said little about mind. He leaves us with an impression, but no exact understanding, of how the evolution of languages (the result of increasing economy in the use of symbols) refines a flow of information into successively more meaningful forms—a hierarchy leading to levels of interpretation manifested as visual perception, natural language, mathematics, and semantic phenomena beyond. Von Neumann was deeply interested in mind. But he wasn't ready to dismantle a concept that could not be reconstructed with the tools available at the time.

Von Neumann's view of the operation of the human nervous system bears more resemblance to the statistically determined behavior of an economic system than to the precisely logical behavior of a digital computer, whether of the 1950s or of today. "The message-system used in the nervous system ... is of an essentially statistical character," he wrote in his Silliman lecture notes, published posthumously in 1958. "In other words, what matters are not the precise positions of definite markers, digits, but the statistical characteristics of their occurrence. . . . Thus the nervous system appears to be using a radically different system of notation from the ones we are familiar with in ordinary arithmetics and mathematics: instead of the precise systems of markers where the position—and presence or absence—of every marker counts decisively in determining the meaning of the message, we have here a system of notations in which the meaning is conveyed by the statistical properties of the message. . . . Clearly, other traits of the (statistical) message could also be used: indeed, the frequency referred to is a property of a single train of pulses whereas every one of the relevant nerves consists of a large number of fibers, each of which transmits numerous trains of pulses. It is, therefore, perfectly plausible that certain (statistical) relationships between such trains of pulses should also transmit information. . . . Whatever language the central nervous system is using, it is characterized by less logical and arithmetical depth than what we are normally used to [and] must structurally be essentially different from those languages to which our common experience refers."[9]

Despite the advances of neurobiology and cognitive science over the past forty years, this fundamental picture of the brain as a mechanism for evolving meaning from statistics has not changed. Higher levels of language produce a coherent residue as this underlying flow of statistical information is processed and refined. Information flow in the brain is pulse-frequency coded, rather than digitally coded as in a computer. The resulting tolerance for error is essential for reliable computation by a web of electrically noisy and chemically sensitive neurons bathed in a saline fluid (or, perhaps, a web of microprocessors bathed in the distractions of the real world). Whether a particular signal is accounted for as excitation or inhibition depends on the individual nature of the synapses that mediate its journey through the net. A two-valued logic, to assume the simplest of possible models, is inherent in the details of the neural architecture—a more robust mechanism than a two-valued code.

Von Neumann's name remains synonymous with serial processing, now implemented by microprocessors adhering to a logical architecture unchanged from that developed at the Institute for Advanced Study in 1946. He was, however, deeply interested in information-processing architectures of a different kind. In 1943, Warren McCulloch and Walter Pitts demonstrated that any computation performed by a network of (idealized) neurons is formally equivalent to some Turing-machine computation that can be performed one step at a time. Von Neumann recognized that in actual practice (of either electronics or biology) combinatorial complexity makes it prohibitively expensive, if not impossible, to keep this correspondence two-way. "Obviously, there is on this level no more profit in the McCulloch–Pitts result," he noted in 1948, discussing the behavior of complicated neural nets. "There is an equivalence between logical principles and their embodiment in a neural network, and while in the simpler cases the principles might furnish a simplified expression of the network, it is quite possible that in cases of extreme complexity the reverse is true."[10]

Von Neumann believed that a complex network formed its own simplest behavioral description; to attempt to describe its behavior using formal logic might be an intractable problem, no matter how much computational horsepower was available for the job. Many years—and many, many millions of artificial-intelligence research dollars later—Stan Ulam asked Gian-Carlo Rota, "What makes you so sure that mathematical logic corresponds to the way we think?"[11] Ulam's question echoed what von Neumann had concluded thirty years earlier, that "a new, essentially logical, theory is called for in order to understand high-complication automata and, in particular, the central nervous system. It may be, however, that in this process logic will have to undergo a pseudomorphosis to neurology to a much greater extent than the reverse."[12] Computers, by the 1980s, had evolved perfect memories, but the memory of the computer industry was short. "If your friends in AI persist in ignoring their past, they will be condemned to repeat it, at a high cost that will be borne by the taxpayers," warned Ulam, who turned out to be right.[13]

For a neural network to perform useful computation, pattern recognition, associative memory, or other functions a system of value must be established, assigning the raw material of meaning on an equitable basis to the individual units of information—whether conveyed by marbles, pulses of electricity, hydraulic fluid, charged ions, or whatever else is communicated among the components of the net. This process corresponds to defining a utility function in game theory or mathematical economics, a problem to which von Neumann and Morgenstern devoted a large portion of their book. Only by such a uniform valuation of internal signals can those configurations that represent solutions to external problems be recognized when some characteristic maximum, minimum, or otherwise identifiable value is evolved. These fundamental tokens coalesce into successively more complex structures conveying more and more information at every stage of the game. In the seventeenth century, Thomas Hobbes referred to these mental particles as "parcels," believing their physical existence to be as demonstrable as that of atoms, using the same logic by which we lead ourselves to believe in the physical existence of "bits" of information today.

Higher-level representations, symbols, abstractions, and perceptions are constructed in a neural network not from solutions arrived at by algorithmic (step-by-step) processing, as in a digital computer, but from the relations between dynamic local maxima and minima generated by a real-time, incomprehensibly complex version of one of von Neumann's games. It is what is known as an n-person game, involving, in our case, a subset of the more than 100 billion neurons, interlaced by trillions of synapses, that populate the brain. Von Neumann and Morgenstern demonstrated how to arrive at reasonable solutions among otherwise hopeless combinatorics by means of a finite but unbounded series of coalitions that progressively simplify the search. A successful, if fleeting, coalition, in our mental universe, may surface to be perceived—and perhaps communicated, via recourse to whatever symbolic channels are open at the time—as an idea. It is a dynamic, relational process, and the notion of a discrete idea or mental object possessed of absolute meaning is fundamentally contradictory, just as the notion of a bit having independent existence is contradictory. Each bit represents the difference between two alternatives, not any one thing at one time.

In a neural network, the flow of information behaves like the flow of currency in an economy. Signals do not convey meaning through encoded symbols; they generate meaning depending on where they come from, where they go, and how frequently they arrive. A dollar is a dollar, whether coming in or going out, and you can choose to spend that same dollar on either gasoline or milk. The output of one neuron can be either debited or credited to another neuron's account, depending on the type of synapse at which it connects. The faint pulses of electric current that flow through a nervous system and the pulses of currency that flow through an economy share a common etymology and a common destiny that continues to unfold. The metaphor has been used both ways. "The currency of our systems is not symbols, but excitation and inhibition," noted D. E. Rumelhart and J. E. McClelland in their introduction to *Parallel Distributed Processing: Explorations in the Microstructure of Cognition,* a collection of papers that focused a revival of neural network research ten years ago.[14] "Each of those little 'wires' in the optic nerve sends messages akin to a bank statement where it tells you how much interest was paid this month," wrote neurophysiologist William Calvin in *The Cerebral Symphony,* a recent tour inside the human mind and brain. "You have to imagine, instead of an eye, a giant bank that is busily mailing out a million statements every second. What maximizes the payout on a single wire? That depends on the bank's rules, and how you play the game."[15]

Raw data generated by the first layer of retinal photoreceptors is refined, through ingenious statistical transformations, into a condensed flow of information conveyed by the optic nerve. This flow of information is then refined, over longer intervals of time, into a representation perceived as vision by the brain. There is no coherent encoding of the image, as generated by a television camera, just a stream of statistics, dealt out, like cards, to the brain. Vision is a game in which the brain bids a continuous series of models and the model that is most successful in matching the next hand wins. Finally, vision is refined into knowledge, and if all goes well, knowledge is condensed into wisdom, over a lifetime, by the mind. These elements of economy associated with the workings of our intelligence are mirrored by elements of intelligence associated with the workings of an economy—a convergence growing more visible as economic processes take electronic form.

This convergence has its origins in the seventeenth century, just as the foundations of electronic logic date back to Hobbes's observation that, given the existence of addition and subtraction, otherwise mindless accounting leads to everything else. "Monies are the sinews of War, and Peace," observed Hobbes in 1642.[16] In his *Leviathan* of 1651 he elaborated: "Mony passeth from Man to Man, within the Commonwealth; and goes round about, Nourishing (as it passeth) every part thereof.... Conduits, and Wayes by which it is conveyed to the

Publique use, are of two sorts; One, that Conveyeth it to the Publique Coffers; The other, that Issueth the same out againe for publique payments. . . . And in this also, the Artificiall Man maintains his resemblance with the Naturall; whose Veins receiving the Bloud from the severall Parts of the Body, carry it to the Heart; where being made Vitall, the Heart by the Arteries sends it out again."[17]

Hobbes drew his analogy with the circulation of the blood, made vital by the heart, not the circulation of an electric fluid in the nerves, made vital by the brain. The galvanic fluid had yet to be identified, and monetary currency had yet to take the leap to abstract, fluid forms. In Hobbes's time, wealth conveyed by the transfer of information rather than substance was rarer than gold. The evolution of financial instruments resembles the evolution of the hierarchies of languages embodied by digital computers; both these developments parallel the evolution, 600 million years earlier, of genetic programs controlling the morphogenesis of multicellular forms of life. A surge of complexity followed. The cellular-programming revolution began in the Cambrian era; the computer software revolution began in the era of von Neumann; the monetary revolution began in the time of Hobbes.

When Hobbes was in Paris during the 1640s, he was joined by a young William Petty, who later founded the science of political economy, relieving Hobbes's ideas of religious controversy and inserting them into the mainstream of economic thought. The son of a Hampshire clothier, Petty (1623–1687) went to sea at age thirteen with only sixpence to his name. After breaking his leg and being put ashore in France, Petty lived by his wits, securing his own education and an introduction to Hobbes, who was keeping a safe distance between himself and the civil war in England until things settled down. Petty assisted Hobbes ("who loved his company") with his treatise on optics, *Tractatus Opticus*, 1644, drawing "Mr. Hobbes Opticall schemes for him," reported Aubrey, "which he was pleased to like."[18] After returning to England, Petty earned a degree in medicine at Oxford in 1649, where, with John Wilkins, he hosted the predecessor of the Royal Society, known as the Philosophical Club. Sir Robert Southwell, matching wits until the end of Petty's life, encouraged his cousin to exercise his mind: "For Intuition of truth may not Relish soe much as Truth that is hunted downe."[19]

Petty became famous in 1650 for saving the life of Anne Green, a servant maid charged with the murder of her premature and evidently stillborn child and sentenced to hang. "After she had suffere'd the law," wrote Aubrey, "she was cut downe, and carried away in order to be anatomiz'd by some yong physitians, but Dr. William Petty finding life in her, would not venter upon her, only so farr as to recover her life."[20] Assisted legally and financially by Petty and his colleagues, Anne Green was married, bore several healthy children, and enjoyed fifteen additional years of life.

In 1655 and 1656, Petty organized the first complete survey of Ireland after it was decided to confiscate the estates of all landowners who could not prove "constant good affection" to the English government. Redistribution of the property, to settle the government's debts, required accurate maps. Although Petty's cartography was impeccable, real and perceived inequities aroused lingering animosities, and in 1660 Petty was challenged by Sir Alan Brodrick to a duel. "Sir William is extremely short sighted," said Aubrey, "and being the challengee it belonged to him to nominate place and weapon. He nominates, for the place, a darke Cellar, and the weapon to be a great Carpenter's Axe. This turned the knight's challenge into Ridicule, and so it came to nought."[21]

Petty had a lifelong interest in naval architecture; he acted as an advisor to Charles II and even predicted the use of auxiliary propulsion in "An attempt to demonstrate that an Engine may be fix'd in a good Ship of 5 or 600 Tonn to give her fresh way at Sea in a calm."[22] He was obsessed with the potential of twin-hulled sailing vessels, of which the prototype was launched in 1662. Petty's third catamaran, christened the *Experiment* by Charles II, was lost with all hands during a severe storm in 1665, a misfortune that failed to sink the idea in Petty's mind. He wrote to Robert Southwell in 1680 that "I have a Treatise ready to Vindicate the designe and the necessity of attempting it, which will make it rise againe when I am dead."[23]

In 1671, Petty produced the first of a series of essays on *Political Arithmetick*, circulated widely in manuscript but published in full only after his death. "It was by him stiled Political Arithmetick," wrote Petty's son Charles in the dedication to the 1690 edition, "in as much as things of Government, and of no less concern and extent, than the Glory of the Prince, and the happiness and greatness of the People, are by the ordinary Rules of Arithmetick, brought into a sort of Demonstration ... where the perplexed and intricate ways of the World, are explained by a very mean piece of Science."[24] Exercising "the art of reasoning by figures upon things relating to government," Petty laid the foundations for a growing association between the powers of numbers and the powers of the state. He thereby helped to establish the statistical basis of *cybernétique*, as the theory of government would be positioned adjacent to the theory of power in the systematic classification of human knowledge formulated a century and a half later by Ampère. In his essay *On the Growth and Encrease and Multiplication of Mankind*, published in extract in 1686, Petty anticipated and countered the arguments made a century later by Malthus. As a founder of the science of political economy, Petty was among the first to methodically examine the origins of wealth.

In 1682, in the brief but precise *Quantulumcunque Concerning Money*, Petty posed the question, "What remedy is there if we have too little Money?" His answer, amplified by the founding of the Bank of England in 1694, would resonate throughout the world: "We must erect a Bank, which well computed; doth almost double the Effect of our coined Money: And we have in England Materials for a Bank which shall furnish Stock enough to drive the Trade of the whole Commercial World."[25] Petty showed that wealth is a function not only of how much money is accumulated, but of the velocity with which the money is moved around. This led to the realization that money, like information but unlike material objects, can, by assuming different forms, be made to exist in more than one place at a single time.

An early embodiment of this principle, preceding the Bank of England by more than five hundred years, was the ancient institution known as tallies—notched wooden sticks issued as receipts for money deposited with the Exchequer for the use of the king. "As a financial instrument and evidence it was at once adaptable, light in weight and small in size, easy to understand and practically incapable of fraud," wrote Hilary Jenkinson in 1911. "By the middle of the twelfth century, there was a well-organized and well-understood system of tally cutting at the Exchequer . . . and the conventions remained unaltered and in continuous use from that time down to the nineteenth century."[26] A precise description was given in 1850 by Alfred Smee (whose speculations on artificial intelligence and neural networks were cited in Chapter 3). To discourage fraud and counterfeiting, Smee developed tamperproof banknotes and the formula for what became known as bank ink—a "magnificent, solid ink . . . used for the Court Minute Books until comparatively recent years."[27] As resident surgeon to the Bank of England and the son of the accountant general, Smee was able to state with authority concerning some tallies preserved as relics that "curiously enough, I have ascertained that no gentleman in the Bank of England recollects the mode of reading them." Tallies are the direct ancestor of digital financial instruments being introduced today.

"The tally-sticks were made of hazel, willow, or alder wood, differing in length according to the sum required to be expressed upon them," explained Smee. "They were roughly squared, and one end was pointed; and on two sides of that extremity, the proper notches, showing the sum for which the tally was a receipt, were cut across the wood. All these operations were performed by the officer called 'the maker of the tallies.' On the other two sides of the instrument were written, also in duplicate, the name of the party paying the money, the account for which it was paid, the part of the United Kingdom to which it referred, and the date of payment; recorded with ink upon the wood, by an officer called 'the writer of the tallies.' When the tally was complete, the stick was cleft length-wise by the maker of the tallies, nearly throughout the whole extent, in such a manner that both pieces retained a copy of the inscription, and one half of every notch cut at the pointed end. One piece was then given to the party who had paid the money, for which it was a sufficient discharge; and the other was preserved in the Exchequer. Rude and simple as was this very ancient method of keeping accounts, it appears to have been completely effectual in preventing both fraud and forgery for a space of seven hundred years. No two sticks could be found so exactly similar, as to admit of being identically matched with each other, when split in the coarse manner of cutting tallies; and certainly no alteration of the particulars expressed by the notches and inscription could remain undiscovered when the two parts were again brought together. And, as if it had been further to prove the superiority of these instruments over writing, two attempts at forgery were reported to have been made on the Exchequer, soon after the disuse of the ancient wooden tallies in 1834."[28]

Exchequer tallies were ordered replaced in 1782 by an "indented cheque receipt," but the Act of Parliament (23 Geo. 3, c. 82) thereby abolishing "several useless, expensive and unnecessary offices" was to take effect only on the death of the incumbent who, being "vigorous," continued to cut tallies until 1826. "After the further statute of 4 and 5 William IV the destruction of the official collection of old tallies was ordered," noted Hilary Jenkinson. "The imprudent zeal with which this order was carried out caused the fire which destroyed the Houses of Parliament in 1834."[29] The notches were of various sizes and shapes corresponding to the tallied amount: a 1.5-inch notch for £1000, a 1-inch notch for £100, a 0.5-inch notch for £20, with smaller notches indicating pounds, shillings, and pence, down to a halfpenny, indicated by a pierced dot. The code was similar to the notches still used to identify the emulsion speed of sheets of photographic film in the dark.

Until the Restoration tallies did not bear interest, but in 1660, on the accession of Charles II, interest-bearing tallies were introduced.

They were accompanied by written orders of loan, which, being made assignable by endorsement, became the first negotiable interest-bearing securities in the English-speaking world. Under pressure of spiraling government expenditures the order of loan was soon joined by an instrument called an order of the Exchequer, drawn not against actual holdings but against future revenue and sold at a discount to the private goldsmith bankers whose hard currency was needed to prop things up. In January 1672, unable to meet its obligations, Charles II declared a stop on the Exchequer. At the expense of the goldsmith bankers, this first experiment with artificial money came to an end.

Money is a medium for communicating value across distances and over time. Tallies represented proof of the value that had been communicated in the form of gold and silver to the king—and a promise that the value would eventually be returned. As the government began to spend increasing amounts of money, more and more tallies remained outstanding among the community of merchants and bullion dealers who had loaned money to the king. Under these circumstances, a market for derivative financial instruments evolved—paper notes issued against the indirect security of tallies rather than the direct security of bullion or coin. Gold and silver that was in the hands of the king or had passed through his hands in being spent was now represented, at the same time, by circulating paper. As long as the paper was sufficiently trusted, and the king didn't issue a stop, it was possible to have one's cake and eat it too—increasing the amount of money in circulation without increasing the amount of gold.

These arrangements evolved, fitfully, into the system of banking and paper currency with which we are familiar today. "A Banke is a certain number of sufficient men of Credit and Estates joyned together in a stock, as it were for keeping several mens Cash in one Treasury, and letting out imaginary money at interest ... to Trades-men or others, that agree with them for the same, and making payment thereof by Assignation, passing each mans Accompte from one to another, yet paying little money," wrote Francis Cradocke in 1660 in his *Expedient For taking away all Impositions, and raising a Revenue without Taxes, By Erecting Bankes for the Encouragement of Trade*.[30] This "imaginary" money sounded too good to be true, and often was. Moving at ever-increasing speed in increasingly abstract, fluid, and intangible forms, it soon came to govern the affairs of human society as a whole. "Above all other Engines or Instruments, the greatest preheminence is due unto a Banck," exclaimed Henry Robinson in 1652. "It is the Elixir or Philosophers Stone, to which all Nations, and every thing within those Nations must be subservient, either by faire meanes or by foule."[31]

An economy is a system that assigns numerical values to tangible and intangible things. These numbers, having a peculiar tendency, common to all numbers, of lending themselves to intelligent processing, start moving the things around. The history of money has been a step-by-step progression from things to numbers: numbers stamped on coins, numbers printed on banknotes, machine-readable codes on checks, coded electronic transfers between numbered accounts, credit-card numbers transferred over the phone, and now a host of competing forms of digital currency, represented by numbers alone. The relations between money and information go both ways: the flow of information conveys and represents money, and the flow of money conveys and represents information. Prices represent the state of relations among different things, and the markets and other mechanisms whereby money, securities, or other abstractions sell at a discount against the future represent predictions of future events.

The cross-fertilization between banking, digital computing, and telecommunications continues a transformation precipitated nine centuries ago when the chamberlains of the Tower of London discovered that by splitting some sticks of wood in two they could double the spending power of their hoard of gold. The same principle that allowed the Exchequer to split a notched piece of wood, assign a value, and rest assured that only one person could turn up with the matching piece of wood demanding to be repaid, is now being implemented in digital form. Nearly all such digital financial instruments—from conventional electronic funds transfers to anonymous digital cash—are based on constructing a very large number with two prime factors that are all but impossible to extract from the product by brute-force attack. The product may be freely revealed, with its factors remaining mathematically concealed, using pairs of public and private cryptographic keys as uniquely mated as the two halves of a split piece of wood, guaranteeing that valuable numbers only have value to those to whom they are legitimately issued or assigned.

The most successful encryption system is known as RSA after its inventors, Ronald Rivest, Adi Shamir, and Leonard Adleman, who introduced it in 1978. "The era of 'electronic mail' may soon be upon us; we must ensure that two important properties of the current 'paper mail' system are preserved: (a) messages are private, and (b) messages can be signed," they jointly announced. "An encryption method is presented with the novel property that publicly revealing an encryption key does not thereby reveal the corresponding decryption key. . . . A message can be 'signed' using a privately held decryption key. Anyone can verify this signature using the corresponding publicly revealed encryption key. Signatures cannot be forged. . . . A message is encrypted by representing it as a number M, raising M to a publicly specified power e, and then taking the remainder when the result is divided by the publicly specified product, n, of two large secret prime numbers p and q. . . . The security of the system rests in part on the difficulty of factoring the published divisor, n."[32]

Unfortunately, cryptographically secure transmission of electronic mail and electronic funds can also be used to hatch terrorist plots, launder ill-gotten gains, and evade taxes by leaving local authorities behind. With good intentions but diminishing success, the United States has tried to keep cryptography under government control. The argument over control of cryptography has been going on since codes began. At the end of his treatise on telecommunications and cryptography published in 1641, Petty's colleague John Wilkins considered the abuse of cryptography by criminal conspirators, concluding that "if it be feared that this Discourse may unhappily advantage others, in such unlawfull courses: Tis considerable, that it does not only teach how to deceive, but consequently also how to discover Delusions. And then besides, the chiefe experiments are of such nature, that they cannot be frequently practiced, without just cause of suspicion, when as it is in the Magistrates power to prevent them. However, it will not follow, that every thing must be supprest, which may bee abused. . . . If all those usefull inventions that are lyable to abuse, should therefore be concealed, there is not any Art or Science, which might be lawfully professed."[33]

Wilkins distinguished between digital coding and pulse-frequency coding, providing an extensive catalog of cryptographic techniques. Noting that "because Words are onely for those that are present both in time and place," he recognized the power of coded information to penetrate barriers not only of distance but of time. He was the first to observe that high-speed data communications would allow the noon price of something in London to be communicated westward before it was noon somewhere else. "Suppose (I say) this Messenger should set forth from London, in the very point of noon," noted Wilkins, envisioning a relay of optical signals, "hee would notwithstanding, arrive at Bristow before twelve of the clock that day. That is, a Message may be by these means conveyed so great a distance, in fewer minutes then those which make the difference between the two Meridians of those places."[34] Time is money. It was the ability to convey market information, trading orders, and fund transfers even slightly ahead of the competition that led to the proliferation of cryptographically secure telecommunications channels by which electronic money has spread throughout the world.

According to a 1995 estimate by the International Telecommunications Union, 2.3 trillion dollars circulates electronically every twenty-four hours—equivalent to 180,000 tons of gold, or a 1,500-mile stack of hundred-dollar bills. Electronic currency has diffused outward from the central banking networks to penetrate the street corner, the desktop, the telephone system, and a host of card-based payment systems, smart and dumb. Banks are becoming networks, and networks are becoming banks. Having seen corporate mainframes replaced with desktop computers, some analysts believe the powers of large banking institutions will be similarly overturned. But the banks are here to stay. "Commercial banking has been around over 600 years," consultant Eric Hughes has explained. "Computers are less than 60 years old. Microcomputer software companies are 20 years old and still reinvent the wheel. Assuming a convergence, who do you think will learn the other's business first?"[35]

The result will be more money, faster money, and money more tightly coupled to things, network architecture, people, and ideas. The scales are shifting both in distance and in time; the intelligence a large corporation once gathered for its annual report is now available to any small business using a personal computer to manage its day-to-day accounts. "We felt that the distinction between micro- and macroeconomics, while appropriate in a non-computer age, was no longer necessary,"[36] remarked economist Gerald Thompson, recalling his final collaboration with Oskar Morgenstern in 1975, two years before Morgenstern's death.

Money is a recursive function, defined, layer upon layer, in terms of itself. The era when you could peel away the layers to reveal a basis in precious metals ended long ago. There's nothing wrong with recursive definitions. (Definition of recursive: see recursive; or, Gregory Bateson's definition of information as "any difference that makes a difference"—the point being that information and meaning are self-referential, not absolute.) But formal systems based on recursive functions, whether in finance or mathematical logic, have certain peculiar properties. Gödel's incompleteness theorems have analogies in the financial universe, where liquidity and value are subject to varying degrees of definability, provability, and truth. Within a given financial system (i.e., a consistent system of values) it is possible to construct financial instruments whose value can be defined and trusted but cannot be proved without assuming new axioms that extend the system's reach. As Gödel demonstrated for logic and arithmetic, there are two sides to this. No financial system can ever be completely secure and closed. On the other hand, like mathematics or any other sufficiently powerful system of languages, there is no limit to the level of concepts that an economy is able to comprehend.

All free-market economies show signs of intelligence, to varying degrees. Conversely, close inspection of many mechanisms we regard as intelligent reveals fundamentally economic systems underneath. When Oskar Morgenstern was asked to explain the power of game theory to a popular audience in 1949, he used the example of a simplified form of poker for two players, using a three-card deck and a one-card, no-draw hand.[37] To determine all possible strategies for this game by brute-force computation requires two billion arithmetic operations. Simple economic systems are able to arrive at practical solutions to problems that are computationally difficult to solve. That brains in nature operate more as economies than as digital computers should come as no surprise. Indeed, economic principles are the only known way to evolve intelligent systems from primitive components that are not intelligent themselves. As Marvin Minsky explained in his *Society of Mind:* "You can build a mind from many little parts, each mindless by itself. . . . Any brain, machine, or other thing that has a mind must be composed of smaller things that cannot think at all."[38] Or, as Samuel Butler put it in 1887: "Man is only a great many amoebas, most of them dreadfully narrow-minded, going up and down the country with their goods and chattels."[39] The archetypal invisible hand of Adam Smith ("He intends only his own gain, and he is in this, as in many other cases, led by an invisible hand to promote an end which was no part of his intention.")[40] appears to be capable of building not only an economy, or a damage-resistant communications network, but a brainlike structure, perhaps a mind. "Probably the closest parallel structure to the Internet is the free market economy," observed Paul Baran.[41]

The incubation of intelligence within a network requires an exceptionally fluid, arborescent structure and the infiltration of this architecture by a statistical language analogous to the primary statistical language that von Neumann identified as the machine language of the brain. At one level, this language may appear to us to be money, especially the new, polymorphous E-money that circulates without reserve at the speed of light. E-money is, after all, simply a consensual definition of "electrons with meaning," allowing other levels of meaning to freely evolve. Composed of discrete yet divisible and liquid units, digital currency resembles the pulse-frequency coding that has proved to be such a rugged and fault-tolerant characteristic of the nervous systems evolved by biology. Frequency-modulated signals that travel through the nerves are associated with chemical messages that are broadcast by diffusion through the fluid that bathes the brain. Money has a twofold nature that encompasses both kinds of behavior: it can be transmitted, like an electrical signal, from one place (or time) to another; or it can be diffused in any number of more chemical, hormonelike ways.

Money has the self-reinforcing tendencies and semantic transparency that allow neural networks to work. The flow of money permeates all components of the network, strengthens frequently used connections, propagates backward, transforms local processing mechanisms, and encourages new connection pathways in response. This architectural plasticity allows neural networks to adapt, remember, and learn to predict events. Freely reversible financial gradients direct when and where new connections are formed and which connections die out. The flow of currency transports, integrates, and accumulates signals; a myriad of financial instruments function as neurotransmitters and bridge synaptic gaps.

"Neural processes are insulated from the extra-cellular fluid by a membrane only approximately 50 angstroms thick," wrote semiconductor pioneer Carver Mead, explaining how to build integrated circuits modeled after the neural circuits found in our brains. "The capacitance of this nerve membrane serves to integrate charge injected into the dendritic tree by synaptic units. Much of the real-time nature of neural computation is vastly simplified because this integrating capability is used as a way of storing information for short periods—from less than 1 millisecond to more than 1 second. There is an important lesson to be learned here, an insight that would not follow naturally from the standard lore of either computer science or electrical engineering. Like the spatial smoothing performed by resistive networks . . . temporal smoothing is an essential and generally useful form of computation."[42] Whether conveyed by bullion or binary numbers, accounts accumulate incoming currency over various periods and release outgoing currency at intervals more or less closely related to patterns generated by the currency coming in. In an age when nanoseconds count, it is easy to forget that the components of a neural net must have some temporal delay, however small, to allow the network to compute.

In drawing these analogies, what of the data that now flood the telecommunications net: pictures, sound, video, interactive data communications, encyclopedias of text? All this traffic means something to somebody, and some of it advances our sciences, our culture, and our arts, but is it the stuff of meaning (or the measure of a utility function) across the system as a whole? Maybe or maybe not. What counts is not so much the data that flow in any given direction, but the money that flows the other way. In the coalescence of the software, banking, and telecommunications industries, we are spawning the precursors of collective digital organisms that will roam the network like social insects, sending packets of digital currency back to their nests. The push toward interactive communications over the Web is aimed not at delivering content to the consumer (this can be done already), but at delivering money, in real time, the other way. Electronic money allows organizations to do things and immediately sense the results.

This was the original premise of purposive systems as expounded by Norbert Wiener and Julian Bigelow in 1943: intelligent behavior evolves as a consequence of the ability to measure and keep account of the effects of a given signal through feedback loops that return a message signifying the magnitude of the result. These principles are common to automatic anti-aircraft guns firing at a moving target, neurons seeking to make the right connections inside a brain, laboratory animals facing a maze, corporations facing a free-market economy, or any other situation where it is possible to place a value on an objective at which to aim.

The goal of life and intelligence, if there is one, is awkward to define. A general aim can be detected in the tendency toward a local decrease in the entropy of that fragment of the universe considered to be intelligent or alive. This is a measurable way of saying that life and intelligence tend to organize themselves. Order, however, is only available in limited quantities, at a certain price. Organization can be increased or created only by absorbing existing sources of order (by eating other creatures as food, joining them in symbiosis, or by photosynthesis exploiting the ordered energy of the sun) or by shedding disorder (by excreting waste, radiating heat, or learning from experience through the attrition of less-meaningful connections in the developing infant brain). In human society, money serves to measure and mediate local markets for decreasing entropy, whether it measures the refinement of an ounce of gold, the energy available in a ton of coal, the price of a share in a multinational organization, or the value of the information accumulated in a book. We invented the science of economics, but economy came first.

In 1965, twenty years after the disbanding of Alan Turing's crew at Bletchley Park, Irving J. Good published his speculations on the development of an ultraintelligent machine, later described as "a machine that believes that people cannot think."[43] Central to the development of an indisputable mechanical intelligence is the question of what meaning is and how meaning is evolved. In Good's analysis, meaning and economy are deeply intertwined; where there is meaning, there is an economy of things representing information (or information representing things) by which the meaning of things can be evaluated and from which meaningful information structures can be built. "The production of meaning can be regarded as the last regeneration stage in the hierarchy," noted Good, "and it performs a function of economy just as all the other stages do. It is possible that this has been frequently overlooked because meaning is associated with the metaphysical nature of consciousness, and one does not readily associate metaphysics with questions of economy. Perhaps there is nothing more important than metaphysics, but, for the construction of an artificial intelligence, it will be necessary to represent meaning in some physical form."[44]

In 1677, William Petty, in a letter to his cousin Robert Southwell on "The Scale of Creatures," wrote that "between God and man, there are holy Angells, Created Intelligences, and subtile materiall beings; as there are between man and the lowest animall a multitude of intermediate natures."[45] Whether he saw economic systems as among these created intelligences is left unsaid. By the time of Alfred Smee, the forest was obscured by the trees. Proposing, in 1851, that mechanical processing of ideas would require a relational and differential machine the size of the City of London, Smee failed to notice from his quarters on Threadneedle Street that the Bank of England's network of linked transactions, mediated by a hive of accountants, already constituted such a machine. "The average daily transactions in the London Bankers' Clearing House amount to about twenty millions of pounds sterling, which if paid in gold coin would weigh about 157 tons," reported Stanley Jevons in 1896.[46]

John von Neumann, although halted in midstream, was working toward a theory of the economy of mind. In the universe according to von Neumann, life and nature are playing a zero-sum game. Physics is the rules. Economics—which von Neumann perceived as closely related to thermodynamics—is the study of how organisms and organizations develop strategies that increase their chances for reward. Von Neumann and Morgenstern showed that the formation of coalitions holds the key, a conclusion to which all observed evidence, including Nils Barricelli's experiments with numerical symbioorganisms, lends support. These coalitions are forged on many levels—between molecules, between cells, between groups of neurons, between individual organisms, between languages, and between ideas. The badge of success is worn most visibly by the members of a species, who constitute an enduring coalition over distance and over time. Species may in turn form coalitions, and, perhaps, biology may form coalitions with geological and atmospheric processes otherwise viewed as being on the side of nature, not on the side of life.

Coalitions, once established, can be maintained across widening gaps, such as the levels of abstraction that separate the metaphysics of a language from the metabolism of its host. Fortunes shift, and if a symbiont develops a strategy that dominates the behavior of its host, the roles may be reversed. Our own species is doing its best to adjust to a three-way coalition of self-reproducing human beings, self-reproducing numbers, and self-reproducing machines. Signs of intelligence are evident at every turn, but because this intelligence envelopes us in all directions the whole picture lies beyond our grasp. We have made only limited progress in the three hundred years since Robert Hooke explained how the soul is somehow "apprehensive" of "a continued Chain of Ideas coyled up in the Repository of the Brain."[47] What mind, if any, will become apprehensive of the great coiling of ideas now under way is not a meaningless question, but it is still too early in the game to expect an answer that is meaningful to us.

10

THERE'S PLENTY OF ROOM AT THE TOP

We're doing this the way you'd plan walkways in a park:
Plant grass, then put sidewalks where the paths form.

—JOE VAN LONE[1]

"There's Plenty of Room at the Bottom" was the title of an after-dinner talk given by physicist Richard Feynman at the California Institute of Technology on 29 December 1959. Feynman's timing was perfect. He kept his audience awake with a series of outlandish speculations that soon turned out to be spectacularly right. "In the year 2000, when they look back at this age," announced Feynman, "they will wonder why it was not until the year 1960 that anybody began seriously to move in this direction." Imagining small machines being instructed to build successively smaller and smaller machines, Feynman estimated the orders of magnitude by which such devices could become cheaper, faster, more numerous, and collectively more powerful. Molecules, and eventually atoms, would supply mass-produced low-cost parts.

"Computing machines are very large; they fill rooms," said Feynman. "Why can't we make them very small, make them of little wires, little elements—and by little, I mean little. For instance, the wires should be 10 or 100 atoms in diameter, and the circuits should be a few thousand angstroms across." Besides all the other good reasons to avoid building computers the size (and cost) of the Pentagon, Feynman pointed out that "information cannot go any faster than the speed of light—so, ultimately, when our computers get faster and faster and more and more elaborate, we will have to make them smaller and smaller.

"How can we make such a device? What kind of manufacturing processes would we use?" Feynman asked. "One possibility we might

173

consider, since we have talked about writing by putting atoms down in a certain arrangement, would be to evaporate the material, then evaporate the insulator next to it. Then, for the next layer, evaporate another position of a wire, another insulator, and so on. So, you simply evaporate until you have a block of stuff which has the elements—coils and condensers, transistors and so on—of exceedingly fine dimensions."[2]

Feynman did not limit his speculations to electronic microprocessors, however intriguing or lucrative these prospects might be, but continued on down to atom-by-atom manufacturing, "something, in principle, that can be done; but, in practice, has not been done because we are too big." He greeted the implications with an enthusiasm "inspired by the biological phenomena in which chemical forces are used in a repetitious fashion to produce all kinds of weird effects (one of which is the author)."[3] He left other even weirder effects unsaid. Many of Feynman's techniques are now in routine use, the convergence between microbiology and microtechnology steadily eroding the underpinnings of distinction between living organisms and machines. No new laws of physics have turned up to render his predictions less probable than they were in 1959.

Yes, there is plenty of room at the bottom—but nature got there first. Life began at the bottom. Microorganisms have had time to settle in; most available ecological niches have long been filled. Many steps higher on the scale, insects have been exploring millimeter-scale engineering and socially distributed intelligence for so long that it would take a concerted effort to catch up. Insects might be reinvented from the top down by the miniaturization of machines, but we are more likely to reinvent them from the bottom up, by recombinant entomology, for the same reasons we are reengineering existing one-celled organisms rather than developing new ones from scratch.

Things are cheaper and faster at the bottom, but it is much less crowded at the top. The size of living organisms has been limited by gravity, chemistry, and the inability to keep anything much larger than a dinosaur under central-nervous-system control. Life on earth made it as far as the blue whale, the giant sequoia, the termite colony, the coral reef—and then we came along. Large systems, in biology as in bureaucracy, are relatively slow. "I find it no easier to picture a completely socialized British Empire or United States," wrote J. B. S. Haldane, "than an elephant turning somersaults or a hippopotamus jumping a hedge."[4]

Life now faces opportunities of unprecedented scale. Microprocessors divide time into imperceptibly fine increments, releasing signals that span distance at the speed of light. Systems communicate globally and endure indefinitely over time. Large, long-lived, yet very fast composite organisms are free from the constraints that have limited biology in the past. Since the process of organizing large complex systems remains mysterious to us, we have referred to these developments as self-organizing systems or self-organizing machines.

Theories of self-organization became fashionable in the 1950s, generating the same excitement (and disappointments) that the "new" science of complexity has generated in recent years. Self-organization appeared to hold the key to natural phenomena such as morphogenesis, epigenesis, and evolution, inviting the deliberate creation of systems that grow and learn. Unifying principles were discovered among organizations ranging from a single cell to the human nervous system to a planetary ecology, with implications for everything in between. All hands joined in. Alan Turing was working on a mathematical model of morphogenesis, theorizing how self-organizing chemical processes might govern the growth of living forms, when his own life came to an end in 1954; John von Neumann died three years later in the midst of developing a theory of self-reproducing machines.

"The adjective [self-organizing] is, if used loosely, ambiguous, and, if used precisely, self-contradictory," observed British neurologist W. Ross Ashby in 1961. "There is a first meaning that is simple and unobjectionable," Ashby explained. "This refers to the system that starts with its parts separate (so that the behavior of each is independent of the others' states) and whose parts then act so that they change towards forming connections of some type. Such a system is 'self-organizing' in the sense that it changes from 'parts separated' to 'parts joined.' An example is the embryo nervous system, which starts with cells having little or no effect on one another, and changes, by the growth of dendrites and formation of synapses, to one in which each part's behavior is very much affected by the other parts."[5] The second type of self-organizing behavior—where interconnected components become organized in a productive or meaningful way—is perplexing to define. In the infant brain, for example, self-organization is achieved less by the growth of new connections and more by allowing meaningless connections to die out. Meaning, however, has to be supplied from outside. Any individual system can only be self-organizing with reference to some other system; this frame of reference may be as complicated as the visible universe or as simple as a single channel of Morse code.

William Ross Ashby (1903–1972) began his career as a psychiatrist, diversifying into neurology by way of pathology after serving in the Royal Medical Corps during World War II. By studying the structure of the human brain and the peculiarities of human behavior, he sought to unravel the mysteries in between. Like von Neumann, he hoped to explain how mind can be so robust yet composed of machinery so frail. Two years before his death, Ashby reported on a series of computer simulations measuring the stability of complex dynamic systems as a function of the degree of interconnection between component parts. The evidence suggested that "all large complex dynamic systems may be expected to show the property of being stable up to a critical level of connectance, and then, as the connectance increases, to go suddenly unstable."[6] Implications range from the origins of schizophrenia to the stability of market economies and the performance of telecommunications webs.

Ashby formulated a concise set of principles of self-organizing systems in 1947, demonstrating "that a machine can be at the same time (a) strictly determinate in its actions, and (b) yet demonstrate a self-induced change of organisation."[7] This work followed an earlier paper on adaptation by trial and error, written in 1943 but delayed by the war, in which he observed that "an outstanding property of the nervous system is that it is self-organizing, i.e., in contact with a new environment the nervous system tends to develop that internal organization which leads to behavior adapted to that environment."[8] Generalizing such behavior so that it was "not in any way restricted to mechanical systems with Newtonian dynamics," Ashby concluded that " 'adaptation by trial and error' . . . is in no way special to living things, that it is an elementary and fundamental property of all matter, and . . . no 'vital' or 'selective' hypothesis is required."[9] Starting from a rigorous definition of the concepts of environment, machine, equilibrium, and adaptation, he developed a simple mathematical model showing how changes in the environment cause a machine to break, that is, to switch to a different equilibrium state. "The development of a nervous system will provide vastly greater opportunities both for the number of breaks available and also for complexity and variety of organization," he wrote. "The difference, from this point of view, is solely one of degree."[10]

When the cybernetics movement took form in the postwar years, Ashby's ideas were folded in. His *Design for a Brain: The Origin of Adaptive Behaviour,* published in 1952, was adopted as one of the central texts in the new field. Ashby's "homeostat," the electromechanical embodiment of his ideas on equilibrium-seeking machines, behaved like a cat that turns over and goes back to sleep when it is disturbed. His "Law of Requisite Variety" held that the complexity of an effective control system corresponds to the complexity of the system under its control.

Ashby believed that the "spontaneous generation of organization" underlying the origins of life and other improbabilities was not the exception but the rule. "Every isolated determinate dynamic system obeying unchanging laws will develop 'organisms' that are adapted to their 'environments,' " he argued. "There is no difficulty, in principle, in developing synthetic organisms as complex and as intelligent as we please. But . . . their intelligence will be an adaptation to, and a specialization towards, their particular environment, with no implication of validity for any other environment such as ours."[11]

According to Ashby, high-speed digital computers offered a bridge between laws and life. "Until recently we have had no experience of systems of medium complexity; either they have been like the watch and the pendulum, and we have found their properties few and trivial, or they have been like the dog and the human being, and we have found their properties so rich and remarkable that we have thought them supernatural. Only in the last few years has the general-purpose computer given us a system rich enough to be interesting yet still simple enough to be understandable . . . it enables us to bridge the enormous conceptual gap from the simple and understandable to the complex." To understand something as complicated as life or intelligence, advised Ashby, we need to retrace its steps. "We can gain a considerable insight into the so-called spontaneous generation of life by just seeing how a somewhat simpler version will appear in a computer," he noted in 1961.[12]

The genesis of life or intelligence within or among computers goes approximately as follows: (1) make things complicated enough, and (2) either wait for something to happen by accident or make something happen by design. The best approach may combine elements of both. "My own guess is that, ultimately, efficient machines having artificial intelligence will consist of a symbiosis of a general-purpose computer together with locally random or partially random networks," concluded Irving J. Good in 1958. "The parts of thinking that we have analyzed completely could be done on the computer. The division would correspond roughly to the division between the conscious and unconscious minds."[13] A random network need not be implemented by a random configuration of neurons, wires, or switches; it can be represented by logical relationships evolved in an ordered matrix of two-state devices if the number of them is large enough. This possibility was inherent in John von Neumann's original conception of the digital computer as an association of discrete logical elements, a population that just so happened to be organized by its central control organ for the performance of arithmetical operations but that could in principle be organized differently, or even be allowed to organize itself. Success at performing arithmetic, however, soon preempted everything else.

An early attempt at invoking large-scale self-organizing processes within a single computer was a project aptly christened Leviathan, developed in the late 1950s at the System Development Corporation, a spin-off division of RAND. Leviathan proposed to capture a behavioral model of a semiautomatic air-defense system that had grown too complicated for any predetermined model to comprehend. Leviathan (the single-computer model) and SAGE (its multiple-computer subject) jointly represented the transition from computer systems designed and organized by engineers to computer systems that were beginning to organize themselves.

The System Development Corporation (SDC) originated in the early 1950s with a series of RAND Corporation studies for the U.S. Air Force on the behavior of complex human-machine systems under stress. In 1951, behind a billiard parlor at Fourth and Broadway in downtown Santa Monica, California, RAND constructed a replica of the Tacoma Air Defense Direction Center, where the behavior of real humans and real machines was studied under simulated enemy attack. The first series of fifty-four experiments took place between February and May 1952, using twenty-eight human subjects provided with eight simulated radar screens under punched-card control. In studying the behavior of their subjects—students hired by the hour from UCLA—it was discovered that participation in the simulations so improved performance that the air force asked RAND to train real air-defense crews instead. "The organization learned its way right out of the experiment," the investigators reported in a summary of the tests. "Within a couple days the college students were maintaining highly effective defense of their area while playing word games and doing homework on the side."[14] The study led to the establishment of a permanent System Research Laboratory within RAND's System Development Division and to a training system duplicated at 150 operational air-defense sites.

RAND's copy of the IAS computer became operational in 1952, followed by delivery of an IBM 701, the first system off the assembly line, in August 1953. The computer systems used to stage RAND's simulations soon became more advanced than the control systems used in actual air defense. "We found that to study an organization in the laboratory we, as experimenters had to become one," wrote Allen

Newell, who went on to become one of the leaders of artificial intelligence research.[15] Repeating the process by which human intelligence may have first evolved, an observational model developed into a system of control. RAND's contracts were extended to include designing as well as simulating the complex information-processing systems needed for air defense. "The simplest way of summarizing the incidents, impressions, and data of the air-defense experiments," reported Newell, "is to say that the four organizations behaved like organisms."[16] RAND's studies were among the first to examine how large information-processing systems not only facilitate the use of computers by human beings but facilitate the use of human beings by machines. As John von Neumann pointed out, "the best we can do is to divide all processes into those things which can be better done by machines and those which can be better done by humans and then invent methods by which to pursue the two."[17]

By the time it became an independent, nonprofit corporation at the end of 1956, the System Development Division employed one thousand people and had grown to twice the size of the rest of RAND. When the air force contracted jointly with the Lincoln Laboratory at MIT and the RAND Corporation to develop the continental air-defense system known as SAGE (Semi-Automatic Ground Environment), the job of programming the system was delegated to SDC. Bell Telephone Laboratories and IBM were offered the contract but both declined. "We couldn't imagine where we could absorb 2,000 programmers at IBM when this job would be over someday," said Robert Crago, "which shows how well we were understanding the future at that time."[18]

SAGE integrated hundreds of channels of information related to air defense, coordinating the tracking and interception of military targets as well as peripheral details, such as some thirty thousand scheduled airline flight paths augmented by all the unscheduled flight plans on file at any given time. Each of some two dozen air-defense sector command centers, housed in windowless buildings protected by six feet of blast-resistant concrete, was based around an AN-FSQ-7 computer (Army–Navy Fixed Special eQuipment) built by IBM. Two identical processors shared 58,000 vacuum tubes, 170,000 diodes, and 3,000 miles of wiring as one ran the active system and the other served as a "warm" backup, running diagnostic routines while standing by to be switched over to full control at any time. These systems weighed more than 250 tons. The computer occupied 20,000 square feet of floor space; input and output equipment consumed another 22,000 square feet. A 3,000-kilowatt power supply and 500 tons of air-conditioning equipment kept the laws of thermodynamics at bay. One hundred air force officers and personnel were on duty at each command center; the system was semiautomatic in that SAGE supplied predigested intelligence to its human operators, who then made the final decisions as to how the available defenses should respond.

The use of one computer by up to one hundred simultaneous operators ushered in the era of time-share computing and opened the door to the age of data networking that followed. The switch from batch processing, when you submit a stack of cards and come back hours or days later to get the results, to real-time computing was sparked by SAGE's demand for instantaneous results. The SAGE computers, descended from the Whirlwind prototype constructed at MIT, also led the switch to magnetic-core memory, storing 8,192 33-bit words, increased to 69,632 words in 1957 as the software grew more complex. The memory units were glass-faced obelisks housing a stack of thirty-six ferrite-core memory planes. It took forty hours of painstaking needlework to thread a single plane; each of its 4,096 ferrite beads was interlaced by fine wires in four directions, the intersections weaving a tapestry of cross-referenced electromagnetic bits. The read-write cycle was six microseconds, shuttling data back and forth nearly 200,000 times from one second to the next. High-speed magnetic drums and 728 individual tape drives supplied peripheral programs and data, and traffic with the network of radar-tracking stations pioneered high-speed (1,300 bits per second) data transmission over the voice telephone system using lines leased from AT&T.

The prototype Cape Cod station was operational in 1953; twenty-three sectors were deployed by 1958; the last six SAGE sector control centers were shut down in January 1984, having outlived all other computers of their time. SAGE was designed to defend against land-based bombers; the age of ballistic missiles left its command centers vulnerable to attack. As the prototype of a real-time global information-processing system, however, SAGE left instructions that are still going strong.

The SAGE operating system incorporated one million lines of code, by far the largest software project of its time. Each control sector was configured differently, yet all sectors had to interact smoothly under stress. To this day no one knows how the system would have behaved in response to a real attack. Even the principal architects of the operating system spoke of it as having been evolved rather than designed. When human beings were added, the behavior of the system became even less predictable, so it was tested regularly with simulated intrusions and mock attacks. The dual-processor configuration allowed these exercises to be conducted using one-half of the system while the other half remained on-line, like the right and left hemispheres of a brain. Exhibiting a quality that some theorists suggest distinguishes organisms from machines, the SAGE system was so complicated that there appeared to be no way to model its behavior more concisely than by putting the system through its paces and observing the results.

Despite the experience with SAGE, in the RAND tradition of giving every hypothesis a chance the Leviathan project was launched. Leviathan was an attempt to let a model design itself. "Leviathan is the name we give to a highly adaptable model of large behavioral systems, designed to be placed in operation on a digital computer," wrote Beatrice and Sydney Rome in a report first published on 29 January 1959.[19] The air force's problem was that its systems were structured and analyzed hierarchically, but when operated under pressure unforeseen relationships caused things to happen between different levels, with unanticipated, and perhaps catastrophic, results. "The problem of system levels ... pervades the investigation of any subject matter that incorporates symbols," wrote the Romes, philosophers by profession and biographers of the seventeenth-century philosopher Nicolas Malebranche. "An example of the latter is any work of art, but the example we shall offer here is drawn from simulating air defense."[20] An oblique reference to "other kinds of systems of command and authority that produce a product or that render a constructive or destructive service" was as close as the Romes came to acknowledging the policy of assured retaliation underlying the air force's interest in decision making by human-machine systems under stress.[21] References to "special Leviathan pushbutton intervention equipment" sound sinister, but only referred to circuits installed at SDC's System Simulation Research Laboratory to allow human operators to input decisions used in training the Leviathan program during tests.

To construct their model, the Romes proposed using a large digital computer as a self-organizing logical network rather than as a data processor proceeding through a sequence of logical steps. "Let us suppose that we decide to use the computer in a more direct, a non-computational way. The binary states of the cores are theoretically subject to change thousands of times each second. If we can somehow induce some percentage of these to enter into processes of dynamic interaction with one another under controllable conditions, then direct simulation may be possible. A million cells of storage subject to rapid individual change may provide a mesh of sufficiently fine grain."[22]

This mesh was to be inoculated with a population of artificial agents corresponding to elements of the reality the model was intended to learn to represent: "The notion of direct representation of real social facts by means of activations in a computing machine can be made clearer if we view the computer employed in such an enterprise as a system of switches, comparable to a large telephone system. . . . Our program, then, begins with a design for an automaton that will be made to operate in a computer. Circuits will be so imposed on the computer that their performance will be analogous to the operations of human agents in real social groups. Next, we shall cut the computer free from its ordinarily employed ability to accomplish mathematical computations. . . . The patterns and entailments of the specific activations will be controlled so that they will be isomorphic with significant aspects of individual human actions. Thus, in a very precise sense, we shall be using a digital computer in an analogue mode. . . . The micro-processes in the computer resemble the micro-functioning of real agents under social constraints."[23] In their 1962 memorandum, the Romes were more specific about the workings of these "artificial agents," even coining a new unit, the "taylor" (after F. W. Taylor, founder of time-and-motion studies), to measure the relative values of four different kinds of "social energy" with which individual agents were endowed.

Then comes the crucial assumption—that without having to pay attention to the details, a behavioral gradient rewarding a closer match between model and reality will encourage the internal structure of the model to organize itself. "As we 'tune' our model, so to speak, that is, make the whole and the components resemble patterns of performance observable in real life, then the adequacy of the specific parametric values imposed on Leviathan for specific experiments should improve; and Leviathan will come to behave more like a real society both in its components and as a whole. In this way system properties can be imposed downward into the hierarchical structures."[24] As the Romes found out, this is easier said than done. They were on the right track, but with Leviathan running on an IBM 7090 computer, even with what was then an enormous amount of memory—over 1 million bits—the track wasn't wide enough. Their model was defeated by the central paradox of artificial intelligence: systems simple enough to be understandable are not complicated enough to behave intelligently; and systems complicated enough to behave intelligently are not simple enough to understand.

Citing von Neumann, the Romes concluded "that neither human thinking nor human social organization resembles a combinational logical or mathematical system."[25] Von Neumann believed the foundations of natural intelligence were distinct from formal logic, but through repeated association of the von Neumann architecture with attempts to formalize intelligence this distinction has been obscured. The Romes, following von Neumann's lead, believed a more promising approach to be the cultivation of these behavioral processes within a random, disorganized computational system, but the available matrix delivered disappointing results. Nonetheless, they foresaw that given a more fertile computational substrate, humans would not only instruct the system but would begin following instructions that were reflected back. "Once we incorporate live agents into our dynamic computer model, the result can be a machine that will be capable of teaching humans how to function better in large man-machine system contexts in which information is being processed and decisions are being made."[26]

The Leviathan project grew into an extensive experiment in communication and information handling in composite human-computer organizations (or, in the Romes' words, organisms), occupying a portion of the twenty-thousand-square-foot Systems Simulation Research Laboratory established at the System Development Corporation's Colorado Avenue headquarters in 1961. Twenty-one experimental human subjects were installed in separate cubicles, equipped with keyboards and video displays connected to SDC's Philco 2000 computer, the first fully transistorized system to be commercially announced. All activity was recorded electronically, and human behavior could be observed by researchers overlooking the complex through one-way glass. The Leviathan Technological System incorporated "16 live group heads (level III) reporting to four live branch leaders ... (level IV) [who] in turn report to a single commanding officer (level V). ... Underneath the live officers ... 64 squads of robots are distributed ... (level II) and report to the group heads directly over them. Each squad of robots consists of artificial enlisted men (level I) who exist only in the computer."[27] The human-machine system was embedded in an artificial environment and fed a flow of artificial work. Under various communication architectures, the Romes observed how organization, adaptive behavior, and knowledge evolved. The Romes concluded that "social hierarchies are no mere aggregations of component individuals but coherent, organic systems in which organic subsystems are integrated. ... Social development is a telic advance, not just a concatenation of correlated events."[28]

With a tightening of air force purse strings, Leviathan was quietly abandoned, but in the SAGE system and its successors (which include most computer systems and data networks in existence today) these principles were given the room to freely grow. They developed not as an isolated experiment, but as an electronic representation coupled as closely to the workings of human culture, industry, and commerce as the SAGE system was coupled to the network of radar stations scanning the continental shelf. SAGE's knowledge of scheduled airline flight paths, for example, gave rise to the SABRE airline-reservation system, refining the grain of the model down to an aircraft-by-aircraft accounting of how many seats are empty and how many are occupied in real time. Ashby's law of requisite variety demands this level of detail in a system that can learn to *control* the routing, frequency, and occupancy of passenger aircraft, rather than simply identifying which flights are passing overhead. The tendency of representative models—whether a bank account that represents the exchange of goods, a nervous system that represents an organism's relation to the world, or a reservation system that represents the number of passengers on a plane—is to translate increasingly detailed knowledge into decision making and control. In less than forty years, adding one subsystem at a time, we have constructed a widely distributed electronic model that is instructing much of the operation of human society, rather than the other way around. The smallest transactions count. Centrally planned economies, violating Ashby's law of requisite variety, have failed not by way of ideology but by lack of attention to detail.

Large information-processing systems have evolved through competition over the limited amount of reality—whether airline customers, market volume, or cold warfare—available to go around. This Darwinian struggle can be implemented within a single computer as well as at levels above; all available evidence indicates that nature's computers are so designed. An influential step in this direction was an elder and much less ambitious cousin of Leviathan named Pandemonium, developed by Oliver Selfridge at the Lincoln Laboratory using an IBM 704. Instead of attempting to comprehend something as diffuse and complex as the SAGE air-defense system, Pandemonium was aimed at comprehending Morse code sent by human operators—a simple but nontrivial problem in pattern recognition that had confounded all machines to date.

Selfridge's program was designed to learn from its mistakes as it went along. Pandemonium—"the uproar of all demons"—sought to embody the Darwinian process whereby information is selectively evolved into perceptions, concepts, and ideas. The prototype operated on four distinct levels, a first approximation to the manifold levels by which a cognitive system makes sense of the data it receives. "At the bottom the data demons serve merely to store and pass on the data. At the next level the computational demons or sub-demons perform certain more or less complicated computations on the data and pass the results of these up to the next level, the cognitive demons who weigh the evidence, as it were. Each cognitive demon computes a shriek, and from all the shrieks the highest level demon of all, the decision demon, merely selects the loudest."[29]

"The scheme sketched is really a natural selection on the processing demons," Selfridge explained. "If they serve a useful function they survive, and perhaps are even the source for other subdemons who are themselves judged on their merits. It is perfectly reasonable to conceive of this taking place on a broad scale—and in fact it takes place almost inevitably. Therefore, instead of having but one Pandemonium we might have some crowd of them. . . . Eliminate the relatively poor and encourage the rest to generate new machines in their own images."[30] In the 1950s, when machine cycles were a rationed commodity and memory cost fifty cents per bit, Pandemonium was far too wasteful to compete with programs in which every instruction, and every memory address, counted. But when parts are cheap and plentiful (whether neurons or microprocessors or object-oriented programming modules), Pandemonium becomes a viable approach. In the 1950s, while computer engineers were preoccupied with architecture and hardware, Selfridge played the outfield in anticipation of semiautonomous processes spawned by machines within machines. Forty years later architecture and hardware are taken for granted and software engineers spend most of their time trying to cultivate incremental adaptations among the host of unseen demons within. "In fact, the ecology of programming is such that overall, programmers spend over 80 percent of their time modifying code, not writing it," said Selfridge in 1995.[31]

It makes no difference whether the demons are strings of bits, sequences of nucleotides, patterns of connection in a random genetic or electronic net, living organisms, cultural institutions, languages, or machines. It also makes no difference whether the decision demon is nature as a whole, an external predator, an internal parasite, a debugging program, or all of the above making a decisive racket at the same time. As Nils Barricelli discovered through his study of numerical symbioorganisms and John von Neumann discovered through his development of game theory, the tendency to form coalitions makes it impossible to keep the levels of decision making from shifting from one moment to the next.

In this ambiguity lies the persistence of the nineteenth century's argument from design—a proprietary dispute over whether the power of selection and the intelligence that it represents belongs to an all-knowing God or to impartial nature alone; whether it descends from above, arises from below, or is shared universally at every level of the scale. "Suppose there were a being who did not judge by mere external appearances," wrote Charles Darwin to Asa Gray in 1857, "but who could study the whole internal organization, who was never capricious, and should go on selecting for one object during millions of generations; who will say what he might not effect?"[32]

The nature of this selective being determines the scale of the effects—and organisms—that can be evolved. Charles Darwin, knowing it best to attempt one revolution at a time, replaced one higher intelligent being with another of a different kind. Intelligence, by any measure, is based on the ability to be selective—to recognize signal amidst noise, to discriminate right from wrong, to select the strategy leading to reward. The process is additive. Darwin was able to replace the supreme intelligence of an all-knowing God, who selected the whole of creation all at once, with the lesser intelligence of a universe that selected nature's creatures step-by-step. But Darwin still dispensed this intelligence downward from the top.

Darwin's success at explaining the origin of species by natural selection may have obscured the workings of evolution in different forms. The Darwinian scenario of a population of individuals competing for finite resources, with natural selection guiding the improvement of a species one increment at a time, tempts us to conclude that where circumstances do not fit this scenario—despite the lengths to which Darwinism has been stretched—evolutionary processes are not involved. Large, self-organizing systems challenge these assumptions—perhaps even the assumption that a system must be self-reproducing, competing against similar systems and facing certain death or possible extinction, to be classified as evolving or alive. It is possible to construct self-preserving systems that grow, evolve, and learn but do not reproduce, compete, or face death in any given amount of time. It is also possible to view large, complex systems, such as species, gene pools, and ecosystems, as information-processing organizations providing a degree of guiding intelligence to component organisms whose evolution is otherwise characterized as blind. A blind watchmaker who can build an eye can evidently assemble structures that no longer stumble around.

Samuel Butler argued against Darwin in 1887 that "we must also have mind and design. The attempt to eliminate intelligence from among the main agencies of the universe has broken down. . . . There is design, or cunning, but it is a cunning not despotically fashioning us from without as a potter fashions his clay, but inhering democratically within the body which is its highest outcome, as life inheres within an animal or plant."[33] Butler did not doubt the power of descent with modification, but he believed Darwin's interpretation of the evidence to be upside down. "Bodily form may be almost regarded as idea and memory in a solidified state—as an accumulation of things each one of them so tenuous as to be practically without material substance. It is as a million pounds formed by accumulated millionths of farthings. . . . The theory that luck is the main means of organic modification is the most absolute denial of God which it is possible for the human mind to conceive—while the view that God is in all His creatures, He in them and they in Him, is only expressed in other words by declaring that the main means of organic modification is, not luck, but cunning."[34] Butler saw each species—indeed, the entire organic kingdom—as a store of knowledge and intelligence transcending the life of its individual members as surely as we transcend the life and intelligence of our component cells.

The dispute between luck and cunning extends beyond evolutionary biology, promising a decisive influence over the future of technology as well. "The notion that no intelligence is involved in biological evolution may prove to be as far from reality as any interpretation could be," argued Nils Barricelli in 1963. "When we submit a human or any other animal for that matter to an intelligence test, it would be rather unusual to claim that the subject is unintelligent on the grounds that no intelligence is required to do the job any single neuron or synapse in its brain is doing. We are all agreed upon the fact that no intelligence is required in order to die when an individual is unable to survive or in order not to reproduce when an individual is unfit to reproduce. But to hold this as an argument against the existence of an intelligence behind the achievements in biological evolution may prove to be one of the most spectacular examples of the kind of misunderstandings which may arise before two alien forms of intelligence become aware of one another."[35] Likewise, to conclude from the failure of individual machines to act intelligently that machines are not intelligent may represent a spectacular misunderstanding of the nature of intelligence among machines.

The step-by-step expression of evolutionary intelligence, compared to the human attention span, is immeasurably slow. The evidence may become inescapable when speeded up. The invisible web of connections that bind an ecology—biological, computational, or both—into a living whole begins to move at a visible pace when the machines evolve from year to year, new generations of software are exchanged in minutes, and control is exercised from one microsecond to the next. Samuel Butler saw that the intelligence of evolution and the intelligence of human beings were heading on a collision course. Butler knew, as did Barricelli, that our definition of intelligence is so anthropocentric as to be next to useless for anything else. "Nothing, we say to ourselves, can have intelligence unless we understand all about it—as though intelligence in all except ourselves meant the power of being understood rather than of understanding," he wrote. "We are intelligent, and no intelligence so different from our own as to baffle our powers of comprehension deserves to be called intelligence at all. The more a thing resembles ourselves, the more it thinks as we do—and thus by implication tells us that we are right, the more intelligent we think it; and the less it thinks as we do, the greater fool it must be; if a substance does not succeed in making it clear that it understands our business, we conclude that it cannot have any business of its own."[36]

As Darwinism was perceived as iconoclasm by the church, so any attempt to attribute intelligence to evolutionary processes is viewed with suspicion by those of orthodox Darwinian faith. Butler and his followers have not stopped at seeking recognition for species-level intelligence, but keep attempting to invoke unorthodox levels of life and mind: "The only thing of which I am sure is, that the distinction between the organic and inorganic is arbitrary; that it is more coherent with our other ideas, and therefore more acceptable, to start with every molecule as a living thing, and then deduce death as the breaking up of an association or corporation, than to start with inanimate molecules and smuggle life into them."[37] Contraband passes across the boundaries between life and nonlife as freely as between intelligence and nonintelligence. The traffic goes both ways, attributing life and intelligence to processes below our own level as well as to those above. Everywhere we look things are turning out to be more intelligent and more alive than they once seemed. The more we come to understand the information-processing systems epitomized by the human brain, the more we find them to be functioning as evolutionary systems, and the more we come to understand evolutionary systems, the more we discover them to be operating as information-processing machines.

Our faith in the intelligence of self-organizing, complex adaptive systems is reminiscent of William Paley's faith in the argument from design. Where we see the emergence of order, Paley saw the hand of design. His protagonist, stumbling upon a watch, finds it utterly impossible "that there existed in things a principle of order, which had disposed the parts of the watch into their present form and situation. He never knew a watch made by the principle of order; nor can he even form to himself an idea of what is meant by a principle of order, distinct from the intelligence of the watchmaker."[38] Comparing watches, we find that Butler's watch was designed from within; Paley's watch was designed from without; Darwin's watch was designed by the accumulation of sheer coincidence over time.

To guard against mechanical error, eighteenth-century navigators carried three chronometers. When one chronometer differed, the other two were assumed to be correct. When you have three watches and no two of them agree, there is no way of telling which, if any, is right. The three differing interpretations of the argument from design given by Butler, Paley, and Darwin correspond loosely to the three different approaches to the design of complex computer systems represented by Leviathan, SAGE, and Pandemonium in 1959. The Romes' Leviathan embodied Butler's faith in nature's mysterious ability to intelligently organize its own design. The air force's SAGE embodied Paley's faith in a centralized, overruling intelligence that administered all instructions from above. Selfridge's Pandemonium embodied Darwin's faith in a tangled bank of subprograms flowering by natural selection out of computational mud.

Over forty years of software development, none of the three approaches has proved entirely wrong. Leviathan attempted a black-box approach to building systems that accumulate empirical knowledge through mechanisms whose details the designer doesn't necessarily understand—a process invoked, if not always admitted, by the authors of most large assemblages of code. SAGE's authoritarian, Grand Central Station approach to system development descended directly, via IBM, to the operating systems that ruled the mainframes of the 1970s and govern the desktops of today. Pandemonium descended, with modification, to the fertility of modular programming and the object-oriented languages and quasi-intelligent agents that are now replicating across the network as a whole. We have transcended the old argument over whether artificial intelligence should be built from linear, sequentially coded processes or incubated within massively parallel webs. Is life the result of linear strings of code-bearing DNA or the result of three-dimensional proteins swimming in auto-catalytic soup? The answer is not one or the other, but both.

The evolution of a diversifying computational ecology from simple strings of 0s and 1s embodies certain ideas in mathematical logic concerning how formal systems evolve into higher types. "I am twisting a logical theorem a little," admitted John von Neumann in a lecture on self-reproducing systems given in December 1949, "a theorem of Gödel that the next logical step, the description of an object, is one class type higher than the object and is therefore asymptotically longer to describe."[39] Evolution is a recursive process, and given the power of recursive functions, we should not be surprised at the complexity and intelligence exhibited by a language or genetic system operating repeatedly on itself. "The possibility of producing an infinite sequence of varieties of descendants from a single program is methodologically significant in a manner which might interest biologists more than artificial intelligencers," remarked logician John Myhill in 1964. "It suggests the possibility of encoding a potentially infinite number of directions to posterity on a finitely long chromosomal tape."[40] In this lies the frustration, and the power, of coded instructions—you cannot always predict the results.

As von Neumann explained to the Hixon Symposium in 1948, "this fact, that complication, as well as organization, below a certain minimum level is degenerative, and beyond that level can become self-supporting and even increasing, will clearly play an important role."[41] This statement is less an echo of Charles Darwin's *Origin of Species* of 1859 than of Robert Chambers's *Vestiges of the Natural History of Creation,* published in 1844: "The idea, then, which I form of the progress of organic life upon the globe—and the hypothesis is applicable to all similar theatres of vital being—is, that the simplest and most primitive type, under a law to which that of like-production is subordinate, gave birth to the type next above it, that this again produced the next higher and so on to the very highest, the stages of advance being in all cases very small." Chambers, writing anonymously and without a scientific reputation to protect, added, "and probably this development upon our planet is but a sample of what has taken place, through the same cause, in all the other countless theatres of being which are suspended in space."[42]

What leads organisms to evolve to higher types? (Darwinian evolution, as Stephen J. Gould, among others, has pointed out, does not "progress" toward greater complexity, but Darwinian evolution, plus symbiogenesis, does.) Is a global electronic intelligence something new, or merely the materialization, on a faster scale, of an intelligence that has existed all along? Natural selection is based on the death, or favored survival, of individuals, and its speed is limited by the time it takes to proceed from one generation to the next. In the age of information the pace of orthodox Darwinism is being left behind. Darwinian evolution, in one of those paradoxes with which life abounds, may become a victim of its own success, unable to keep up with non-Darwinian processes that it has spawned. Erasmus Darwin may turn out to be right.

We have been moving in this direction for some time. "Cultural patterns are in a sense a solution of the problem of having a form of inheritance which doesn't require killing of individuals in order to evolve," observed Nils Barricelli in 1966. "You can evolve them by selecting for cultural patterns, and in this respect it would be a much faster evolutionary phenomenon."[43] The same goes for digital organisms, which do not need to die in order to evolve, although, if memory is limited, the threat of death may help. It also applies to biochemical circuits, such as the molecular hypercycles that preceded the origin of life, or to the topology of an electronic network—a pattern of connections that persists over time, transcending the individual lifetimes of the components from which it is formed. Individual cells are persistent patterns composed of molecules that come and go; organisms are persistent patterns composed of individual cells that come and go; species are persistent patterns composed of individuals that come and go. Machines, as Butler showed with his analysis of vapor engines in 1863, are enduring patterns composed of parts that are replaced from time to time and reproduced from one generation to the next. A global organism—and a global intelligence—is the next logical type, whether we agree with the diagnosis, the terminology, or the assumption of life and intelligence or not.

"I have been trying to think of the earth as a kind of organism, but it is no go," wrote physician Lewis Thomas in 1971. "I cannot think of it this way. It is too big, too complex, with too many working parts lacking visible connections. The other night, driving through a hilly, wooded part of southern New England, I wondered about this. If not like an organism, what is it like, what is it most like? Then, satisfactorily for that moment, it came to me: it is most like a single cell."[44] What appeared to be a single cell in 1971 appears to be something more than a single cell today. Contemplating an ant colony, Thomas wrote that "you begin to see the whole beast, and now you observe it thinking, planning, calculating. It is an intelligence, a kind of live computer, with crawling bits for its wits."[45] Comparing human beings to the ants, Thomas observed that "we are linked in circuits for the storage, processing, and retrieval of information, since this appears to be the most basic and universal of all human enterprises. It may be our biological function to build a certain kind of Hill."[46] With computer networks still in a tenuous, experimental stage, it was nonetheless obvious to Lewis Thomas that "all 3 billion of us are being connected by telephones, radios, television sets, airplanes, satellites, harangues on public-address systems, newspapers, magazines, leaflets dropped from great heights, words got in edgewise. We are becoming a grid, a circuitry around the earth. If we keep at it, we will become a computer to end all computers, capable of fusing all the thoughts of the world."[47]

As a physician and biologist, Lewis Thomas placed the health of human beings and their fellow creatures first. "The most profoundly depressing of all ideas about the future of the human species is the concept of artificial intelligence," he later wrote. "That these devices will take over and run the place for human betterment or perhaps, later on, for machine betterment, strikes me as wrong in a deep sense, maybe even evil. . . . Machines like this would be connected to each other in a network all around the earth, doing all the thinking, maybe even worrying nervously. But being right all the time."[48]

Thomas's caution is understandable. On another level, the level of an earth that "seen from the right distance, from the corner of the eye of an extraterrestrial visitor . . . must surely seem a single creature, clinging to the round warm stone, turning in the sun,"[49] his concerns may be misplaced, or at least no more unsettling than the tyranny of any nervous system over an individual organism's component cells. As Nils Barricelli demonstrated with the growth of his numerical symbioorganisms at the Institute for Advanced Study in 1953, safety in numbers is a fact of life—perhaps *the* fact of life from which all other facts of life evolved. "An organic being is a microcosm," observed Charles Darwin in 1868, "a little universe, formed of a host of self-propagating organisms, inconceivably minute, and numerous as the stars."[50]

From atoms through metazoans to spiral galaxies, the laws of nature form a hierarchy we have yet to comprehend. The ten million transistors now engraved on a single square centimeter of our earth share their digital lifeblood with billions of fellow microprocessors, weaving a fibrous cocoon that spans the globe with a web of light. "If you take a cubic foot of sea water, you might very well find a small flounder in it," Philip Morrison pointed out. "That is hopelessly far from the steady-state. . . . In a cubic mile you could find a submarine full of crew members and software, a still more complex configuration."[51] This hierarchy extends in both directions: in a cubic centimeter you might find a protozoan; in a cubic astronomical unit you might find a single collective organism, clinging to a warm planet, turning slowly in the sun.

11

LAST AND FIRST MEN

There evolved at length a very different kind of complex organism, in which material contact of parts was not necessary either to coordination of behaviour or unity of consciousness. . . . Such was the single-minded Martian host which invaded the Earth.

—OLAF STAPLEDON[1]

Christmas 1917 was the fourth Christmas celebrated under the shadow of World War I. The winter, if not as severe as that of 1916–1917, was still cold enough to freeze mud, machines, and human flesh. William Olaf Stapledon (1886–1950), an English ambulance driver attached to the Sixteenth Division of the French infantry, wrote to his Australian cousin Agnes Miller on 23 December, "two or three people have to take it in turns to grind away at the starting handle and apply hot cloths to the induction pipe for half an hour or more before she will fire at all."[2] World War I, among its other distinctions, marked the transition of modern warfare from horses to machines. Over one-lane roads, horses set the pace. During the Champagne offensive in April 1917, Stapledon was following a galloping artillery team across an exposed stretch of road to retrieve casualties from the front when a shell landed just in front of his ambulance "and the road was immediately blocked with a confusion of splintered wood and the bodies of horses and men."[3]

The first task of the ambulance driver was to distinguish the living from the dead. Those who survived the shells, the bombs, the bullets, the hand grenades, and the gas were fortunate to have volunteers such as Olaf Stapledon to evacuate them to makeshift facilities away from the madness of the front. A pacifist but not a Quaker, Stapledon joined the Friends' Ambulance Unit in 1915 when the operation was in dire need of help. "The Friends' Ambulance Unit, an organisation of young Quakers who wished to carry on the great tradition of their faith by serving the wounded under fire while

193

refusing to bear arms or submit to military discipline . . . sounded like the real thing," Stapledon later explained. "It also offered a quick route to the front."[4]

Stapledon received five weeks of training, learning to drive while studying the rudiments of mechanical and medical first aid. "My brain is full of 'sparking plugs,' 'gudgeon pins,' 'carburetors,' 'exhaust valves,' 'clutches,' & 'throttles,' " he wrote in March 1915. "Unfortunately it is also full of 'scapula,' 'fibula,' 'complicated fractures,' 'spinal columns,' and 'femurs,' and I begin to forget which are human and which mechanical."[5] Visits to the emergency ward of the Liverpool hospital provided a glimpse of things to come. Stapledon was familiar with the dispensary's working-class clientele; as a tutor for the Workers' Education Association he had delivered his first series of lectures, on the history of industrialism, to an audience of Liverpool dockyard workers in 1912. According to Stapledon, his students taught him more than he taught them. He joined the ambulance unit deeply troubled at the burdens of war, like those of industry, being borne by those who stood to benefit the least. Which side would win the war remained unclear; it was certain from the beginning that the working class, on both sides, would lose the most.

Doctors removed a troublesome appendix, Stapledon's father donated a custom-built Lanchester motor ambulance, and Stapledon was off to the quagmire of the western front. Ambulance duty was plagued with ambiguities. Quaker doctrine forbade submission to military discipline, yet to gain access to the front lines the ambulance units had to submit to military control. Assistance was to be given only to the wounded, but wasn't this helping those who might fight again? When Stapledon stopped to clear the road of the gun crew blown to bits a few steps ahead, was he expediting the evacuation of the wounded, according to Quaker principles, or transgressing those principles by clearing the way for the gun crew following behind? As the war dragged on and compromise with the military authorities grew more defined, many left the ambulance unit to join their compatriots in the trenches or returned to England as conscientious objectors to offer witness against the war by going to jail. Stapledon stuck it out to the end. The final months of the war left even Stapledon, advancing behind the French army through the devastation of no-man's-land, at a loss for words. After suffering through years of enemy assaults, Stapledon found it no less difficult to witness the loss of life and limb inflicted by his own side.

Stapledon's convoy, known as Section Sanitaire Anglaise Treize, or S.S.A. 13, reached a full strength of twenty ambulances and forty-five men. Between February 1914 and January 1919 they transported 74,501 patients over 599,410 kilometers of evacuation runs.[6] The Friends' Ambulance Service lost a total of twenty-one members during the war, receiving citations for bravery under fire numerous times. Stapledon was decorated with the *croix de guerre*, but apart from a hernia suffered by hand-cranking a cold engine he survived the war unharmed. "Yes, it was an attempt to have the cake and eat it, to go to war and be a pacifist," he afterward confessed. "Its basis was illogical; but it was a sincere expression of two overmastering and wholesome impulses, the will to share in the common ordeal and the will to make some kind of protest against the common folly."[7] Ambulance drivers saw the worst of war's results. For every life they saved, they ministered as best they could to others they were powerless to help. "Last night as I was going to sleep in my car I thought of the last person who had lain where I was lying," wrote Stapledon in October 1918, three years of bloodshed having failed to dull his anguish over every passenger he lost. "I was perplexed whether to go slow to save him pain, or fast to save his life."[8] Alternating between the terror of battle and the boredom that intervened, Stapledon somehow found it possible "to catch a surprising glimpse of a kind of superhuman beauty in the hideous disaster of war itself."[9]

Peaceful moments intruded here and there. "The moon is brilliant, and the earth is a snowy brilliance under the moon. Jupiter, who was last night beside the moon, is now left a little way behind. Venus has just sunk ruddy in the West, after being for a long while a dazzling white splendour in the sky," wrote Stapledon on Boxing Day, 26 December 1917, after a Christmas whose exhausted spirit had penetrated the defenses on both sides. "I have just come in from a walk with our Professor [Lewis Richardson], and he has led my staggering mind through mazes and mysteries of the truth about atoms and electrons and about that most elusive of God's creatures, the ether. And all the while we were creeping across a wide white valley and up a pine clad ridge, and everywhere the snow crystals sparkled under our feet, flashing and vanishing mysteriously like our own fleeting inklings of the truth about electrons. The snow was very dry and powdery under foot, and beneath that soft white blanket was the bumpy frozen mud. The pine trees stood in black ranks watching us from the hill crest, and the faintest of faint breezes whispered among them as we drew near. The old Prof (he is only about thirty-five, and active, but of a senior cast of mind) won't walk fast, and I was very cold in spite of my sheepskin coat; but after a while I grew so absorbed in his talk that I forgot even my frozen ears. . . . We crossed the ridge through a narrow cleft and laid bare a whole new land, white as the last, and bleaker. And over the new skyline lay our old haunts and the lines. Sound of very distant gunfire muttered to us. . . . What a night it is."[10]

"Professor" Lewis Fry Richardson, meteorologist and mathematical physicist (see Chapter 5), was of solid Quaker background and sought to join the Friends' Ambulance Unit when it was formed in 1914. Two years later he was assigned to Stapledon's convoy after a protracted delay securing leave from his job. He was serving as superintendent of the Meteorological and Magnetic Observatory at Eskdalemuir in Dumfriesshire, Scotland, a branch of the National Physical Laboratory that moved to Eskdalemuir from Kew, near London, when electric railways came into use. The new location was deliberately situated as far as possible from any artificial magnetic fields.

The damp, secluded outpost suited Richardson, who cultivated an "intentionally guided dreaming" that thrived in isolation. "If the machine ran entirely of its own accord, one would have no control, and the dream would be out of touch with reality," he explained. "It is the 'almost' condition that is advantageous for creative thinking. . . . In some ways it is a nuisance; for example I am a bad listener because I am distracted by thoughts; and I was a bad motor-driver because at times I saw my dream instead of the traffic."[11]

Richardson lifted the spirits of the convoy with his peculiar resistance to the hardship, exhaustion, and boredom—occasionally interrupted by shrapnel—that was in the air. "This billet is a barn like the last, but we are far more crowded together," wrote Stapledon on 12 January 1918. "Beside me sits Richardson, the 'Prof,' setting out on an evening of mathematical calculations, with his ears blocked with patent sound deadeners."[12] Richardson was engaged in an extended computation whose character was ideally suited to the long-drawn-out war. One step at a time he was constructing a cellular, numerical model of the weather over northwest Europe for a single six-hour period long past, treating the motions of the atmosphere as satisfying a system of differential equations that linked the conditions in adjacent cells from one interval to the next. This project combined the two contributions to mathematical physics that Richardson had worked on before the war: a theory of eddy diffusion and a method of finite differences to calculate approximate solutions to systems of differential equations resistant to analytic approach. Richardson's mathematics was sound and the historical observations with which he seeded his model were accurate as far as they went, but the prediction that resulted was completely at odds with what had actually happened to the weather over Germany on 20 May 1910.

After the war Richardson published the details of his calculations in a thin but far-reaching volume, *Weather Prediction by Numerical Process*, so that others might learn from his mistakes. He estimated that sixty-four thousand human computers, working under conditions better than those of an unheated barn, could collectively calculate a global model of the atmosphere faster than the weather could keep up. "Imagine a large hall like a theatre, except that the circles and galleries go right round through the space usually occupied by the stage," he wrote, describing the vision that had preoccupied him during the war. "The walls of this chamber are painted to form a map of the globe. The ceiling represents the north polar regions, England is in the gallery, the tropics in the upper circle, Australia on the dress circle and the Antarctic in the pit. A myriad computers are at work upon the weather of the part of the map where each sits, but each computer attends only to one equation or part of an equation. The work of each region is coordinated by an official of higher rank. Numerous little 'night signs' display the instantaneous values so that neighbouring computers can read them. Each number is thus displayed in three adjacent zones so as to maintain communication to the North and South on the map. From the floor of the pit a tall pillar rises to half the height of the hall. It carries a large pulpit on its top. In this sits the man in charge of the whole theatre; he is surrounded by several assistants and messengers. One of his duties is to maintain a uniform speed of progress in all parts of the globe."[13]

Each individual atmospheric cell and its delegated squad of human computers communicate only with its immediate neighbors, yet the combined result is a model of the atmosphere as a whole. Simple, local rules produce complicated, global results. Richardson's methods are identical with the way in which massively parallel computation, distributed among multiple processors, is now used to simulate complex physical systems and similar to the way in which biological intelligence results from the collective effort of large numbers of components communicating less intelligently among themselves.

We do not know how much of this was revealed to Stapledon, but we do know that "discussing the universe," as Stapledon put it, was one way he and Richardson passed their off-duty time. Stapledon learned about electrons, electromagnetic fields, and other highlights of theoretical physics from Richardson, who added traces of electrical engineering here and there. "The other day my electric lighting dynamo went wrong," noted Stapledon on 8 December 1916, not long after Richardson arrived. "The mechanic was away, & I know little about electricity, so I was dished. Fortunately we found that our eccentric meteorologist was also an expert electrician. He and I had a morning on the job unscrewing, tinkering, cleaning and generally titivating, sometimes lying under the car in the mud, sometimes strangling ourselves among machinery inside."[14]

Richardson had spent the years 1909–1912 as laboratory director for the Sunbeam Lamp Company, after gaining his familiarity with electrons directly from the source. He had arrived at King's College, Cambridge, in 1900 to study physics under Cavendish Professor Joseph John Thomson (1856–1940), who had discovered the electron three years before. By measuring the deflection of a stream of charged particles under the influence of electric and magnetic fields, Thomson determined that the charge was about the same as that of a charged hydrogen atom, but carried by particles with about one two-thousandth of the mass. Thus began the age of electronics and a cascade of discoveries precipitated by this demonstration that atoms were not indivisible, but composed of subsidiary parts. Even more remarkably—and how this sparkled in the December moonlight we can only guess—these electrons were found to behave under certain conditions as if they possessed a mind of their own.

"An electron within an atom, they say, has no distinct individuality at all," Stapledon would write thirty years later in a short novel titled *Death into Life*. "It is a mere factor pervading the whole atom. So, equally, with these disembodied individual spirits, dissolved in corporate beings. But the electron may recover its individuality and leap free from the atom, to join perhaps with some other atom and once more die from individuality into a new corporate being. So also with these spirits."[15] By this time Stapledon was an acclaimed writer of science fiction, though he rarely used that term himself.

Stapledon returned from France at the end of January 1919, driving from Dover to Liverpool in a beat-up Sunbeam motorcar that had been donated to the unit by his father and, unlike the ambulances, was not wanted by the Boulogne detachment of the British Red Cross. He married his cousin Agnes Miller (1894–1984) in July 1919, taking her and the Sunbeam on an extended honeymoon in the Lake District before settling down, in a house purchased by his generous shipping-agent father, to raise two children, teach sporadically, and write. His reputation was secured by his first novel, *Last and First Men*, published in 1930, during that dark period when the wounds of World War I were scarcely healed yet the first signs of a sequel were in the air. A gloomy assessment of the human condition pervaded all of Stapledon's fiction, offset by a faith in the possibility of transcending a divisive nature through the evolution of communal mind. A committed socialist, Stapledon leaned toward communism for most of his life but recoiled from the injustices executed in its name. Through the agency of distributed, communal intelligence, Stapledon envisioned the ideals of communism without the dangers of centralized control.

Stapledon's novel, subtitled *A Story of the Near and Far Future*, chronicled life's expression throughout the solar system, looking two billion years ahead. Seven years later, *Star Maker* expanded the field of view to cosmological scale. Stapledon touched on themes that would sustain both science and science fiction for the next sixty years—from artificial intelligence, genetic engineering, and atomic energy to extraterrestrial beings, interstellar transport, and the gradual expansion of civilization into a loosely organized, light-gathering veil around its sun. These two books pursue a quest for meaning through the heavens and beyond. "Was man indeed, as he sometimes desired to be, the growing point of the cosmical spirit, in its temporal aspect at least? Or was he one of many million growing points? Or was mankind of no more importance in the universal view than rats in a cathedral?"[16]

Last and First Men presents a closing history of our species, reviewed by one of our descendants as stellar catastrophe is bringing our solar system to an end. Humanity rises and falls through a succession of mental and physical transformations, regenerating after natural and artificial disasters and emerging in the end into a polymorphous group intelligence, a telepathically linked community of ten million minds spanning the orbits of the outer planets and breaking the bounds of individual consciousness, yet still incapable of more than "a fledgling's knowledge" of the whole. Some ten million years into the story, a virulent species of Martian intelligence invades the earth. Navigating through space on the solar wind, clouds of Martian microorganisms arrive in search of a source of water for their dehydrated home. Individually powerless both in body and in mind, the Martian "sub-vital units" maintain communication via faint electromagnetic fields. When gathered together by the billions they constitute a collective intelligence, "something between an extremely well-disciplined army of specialized units, and a body possessed by one mind."[17]

"Terrestrial organisms, and Martian organisms of the terrestrial type, maintained themselves as vital units by means of nervous systems, or other forms of material contact between parts," Stapledon explained. "In the most developed form, an immensely complicated neural 'telephone' system connected every part of the body with a vast central exchange, the brain. Thus on the earth a single organism was without exception a continuous system of matter, which maintained a certain consistency of form. But . . . Martian life developed at length the capacity of maintaining vital organization as a single conscious individual without continuity of living matter. . . . The Martian organism depended, so to speak, not on 'telephone' wires, but on an immense crowd of mobile 'wireless stations,' transmitting and receiving different wavelengths according to their function. The radiation of a single unit was of course very feeble; but a great system of units could maintain contact with its wandering parts."[18]

The Martian particles were able to aggregate faint magnetic forces to exert measurable physical strength. "Thus a system of materially disconnected units had a certain cohesion," the historian explained. "Its consistency was something between a smoke-cloud and a very tenuous jelly."[19] Faced with attacks by humans and disruption by indiscriminately propagated electromagnetic radiation, the Martians learned to take defensive, and later offensive, forms. It took the two species a long time to recognize each other as intelligent: humans were unable to see anything intelligent in the diffuse Martian clouds; Martians were unable to see anything intelligent in the mute, gravity-bound individuals who appeared to have no mind. In our wireless stations the Martians discerned crude attempts at communication but could find no evidence of organized mentality, concluding that "their only feat seemed to be that they had managed to get control of the unconscious bipeds who tended them."[20]

After fifty thousand years and repeated invasions the Martians succeeded in establishing a permanent colony on the earth. The two races, Martian and human, reached an uneasy coexistence until decadence among the humans led to weaknesses the Martians could not resist. A final war ensued. As the battle dragged on, human scientists invented a bacterium fatal to the Martian organisms but with side effects known to kill people as well. Our politicians, after brief discussions, decided this dangerous pesticide should be used. The Martian society disintegrated, carrying the epidemic back to their home planet, while only a few pockets of shattered humanity survived, plagued for millions of years by an infection of Martian "sub-vital units" that had learned to survive like viruses in the bodies of animals and humans. These microorganisms retained their electromagnetic tendencies, and over the course of vast ages of biological time, "certain species of mammals so readjusted themselves that the

Martian virus became not only harmless but necessary to their well-being. A relationship which was originally that of parasite and host became in time a true symbiosis, a co-operative partnership, in which terrestrial animals gained something of the unique attributes of the vanished Martian organisms. The time was to come when Man himself should look with envy on these creatures, and finally make use of the Martian 'virus' for his own enrichment."[21] Thus telepathic abilities were realized among humankind.

The word *telepathy* was coined by Frederic W. H. Myers (1843–1901), an English poet and school inspector who founded the Society for Psychical Research in 1882. Of vigorous mental and physical constitution, Myers became, at the age of twenty-one, the first Englishman to swim across the channel below Niagara Falls. Claiming that "experiment proves that telepathy—the supersensory transference of thoughts and feelings from one mind to another—is a fact," he attracted a substantial late-Victorian following, especially as new discoveries in physics opened avenues of possibility to otherwise implausible ideas.[22] J. J. Thomson, among other reputable scientists who were asked to witness Myers's experiments, kept an open mind, while making it clear that at least two of the demonstrations he attended showed evidence of fraud.

Myers's eldest son, Leopold (1881–1944), was a promising novelist (*The Orissers*, 1923; *The Near and the Far*, 1929; *The Root and the Flower*, 1935) who never escaped the shadow of his father and never adjusted to the pretensions of literary success. There was little adventure in his life. While Stapledon was rescuing the wounded during the Great War, Leo held a position at the Board of Trade. In 1901, on the death of his father, Leo accompanied his mother to the United States, where a posthumous communication with her husband had been prearranged. This final experiment failed, and Leo was on his own. Drawn by admiration for *Last and First Men*, he became Stapledon's closest literary friend and correspondent until he committed suicide in 1944. Frederic Myers contributed the idea of telepathy that infused Stapledon's books; Leo Myers contributed encouragement as Stapledon developed a literary career. Stapledon's writings echo the proclamation made by the elder Myers in his *Science and a Future Life:* "We ourselves are a part, not only of the race, but of the universe. It is conceivable that our share in its fortunes may be more abiding than we know; that our evolution may be not planetary but cosmical, and our destiny without an end."[23]

Preferring the exotic to the occult, Stapledon went to great lengths to give telepathy a physically explainable form. Technology, however, was about to achieve the same objective, without extraterrestrial help. Millions of years before our Martian adventures were set to unfold, we have been invaded by subvital units—microprocessors—not from the sands of Mars, but from the sands of Earth. Silicon and oxygen, forged from helium and hydrogen in the atomic furnaces of stars, have lingered as the two most common elements in the outer layer of the planet we call home. One atom of silicon combines with two atoms of oxygen to produce silicon dioxide, or silica, which makes up 59 percent of the thin, floating crust—a silica wafer—that is solid ground to us. Silica, in one form or another, is the principal ingredient of 95 percent of the rock beneath our feet.

Exobiologists consider silicon a possible platform for extraterrestrial life. On our planet, carbon-based life came first, although, according to the theories of A. G. Cairns-Smith, siliceous clays may have given our genetic system its start. Self-reproducing clay crystals may have served as a template for the beginnings of organic life, just as organic life is now serving as a substrate for the proliferation of self-reproducing forms of silicon and their associated code. But the development of silicon-based cyberplasm, governing the course of organic life, does not mean that carbon-based metabolism will be superseded or replaced. The distinction between chemistry and electronics appears clear-cut from a distance, but as you look closer the separation is less distinct. The fabrication of microelectronic components is largely a chemical process, while the fabrication of chemical structures depends largely on relations between electrons. We shall not see biochemistry replaced by electronics; we shall see a merger that incorporates them both.

Evolution is traditionally portrayed as a succession of discrete layers, those of geology and biology gathered into chapters like the pages of a book. A layer of dinosaurs is followed by a layer of mammals. But the precursors of mammals were there all along. If you could ask the mammals how long they had been around, they wouldn't answer, "Since the dinosaurs left," they would answer, "Since life began." If you could ask the same question of machines such as microprocessors, you might get an answer that begins not with the age of computers but with the age of bifacial stones. When silica is liquefied by heat and pressure within the earth, then ejected toward the surface and cooled, the result is a glassy substance, obsidian or flint. Glass is a brittle, noncrystalline material that breaks with a characteristic conchoidal fracture, leaving an edge that is scalpel sharp.

For billions of years glass lay scattered by occasional cataclysms about the earth until, roughly two million years ago, it attracted the attention of humanoid hands and minds. Our hands discovered the killing, dismembering power of the edge, while mind discovered the power of selecting, and later duplicating, certain shapes that were found to work the best. Through a convoluted, coevolutionary process, fragments of silica began to be reproduced. The human presence selectively favored increasingly complex chips of silica, given form by the force of information, while the powers gained through the manufacture of these artifacts favored increasingly complex constructs of the human mind.

Silicon, however, had at least two more innings in the game. Pure, crystalline silicon—or rather, pure silicon rendered impure in exactly the right way—was discovered to be a semiconductor, able to act as an electrical switch with electrons as its only moving parts. The first practical application was the silicon-crystal detector for wireless signals, developed by G. W. Pickard in 1906, followed many years later by the diode, the transistor, and then the multiple-transistor integrated circuit, which, if the transistor is considered a Paleolithic implement, compares to the Neolithic, multiple-fluted Clovis point. Then came the microprocessor, a working copy of the machine Julian Bigelow had built at the Institute for Advanced Study twenty years earlier imprinted on a single chip. The two-thousand transistor Intel 4004 appeared in 1971, the one-million transistor Intel 486 in 1990, the five-million transistor Pentium Pro in 1996, with ten-million-transistor digital-signal processors now being introduced. These fragments of silicon, chiseled on submicron scale, depend, like Stapledon's subvital units, on communication for collective strength. Once again silica intervened, this time as thin strands of quartz, forged not into discrete devices such as microblades or microchips, but woven into a fiber-optic network spreading across the surface of the earth like a spider's web.

Telepathic communication in most of Stapledon's societies led to a distributed communications network whereby information was shared freely throughout the species without producing a level of consciousness higher than that of the underlying minds. In other societies, such as the "18th Men," individuals were "capable of becoming on some occasions mere nodes in a system of radiation which itself should then constitute the physical basis of a single mind."[24] The difference depends on bandwidth: the amount of information that can be conveyed over a given communication channel in a given time. If bandwidth begins to match the internal processing power of the individual nodes in a communications network, individuality begins to merge. Since human beings (usually) think at a higher speed than that at which they communicate, this situation does not occur in human society, though hints of it may be found in certain ritual activities in which thoughts are synchronized through music or dance and communication between individuals is speeded up. We can only speculate how mind might emerge, and perhaps has emerged, among aquatic creatures able to visualize *and* telecommunicate by means of underwater sound. Or among millions-of-instructions-per-second microprocessors linked by millions-of-bits-per-second fibers into a telecommunicating web.

Astronomer Fred Hoyle, whose science fiction novel *The Black Cloud* was inspired by Stapledon's ideas, let two of his characters explain as they begin to realize that an electromagnetic, loosely distributed intelligence of interstellar origin is paying a visit to the vicinity of Earth:

" 'The volume of information that can be transmitted radiatively is enormously greater than the amount that we can communicate by ordinary sound. We've seen that with our pulsed radio transmitters. So if this Cloud contains separate individuals, the individuals must be able to communicate on a vastly more detailed scale than we can. What we can get across in an hour of talk they might get across in a hundredth of a second.' "

" 'Ah, I begin to see light,' broke in McNeil. 'If communication occurs on such a scale then it becomes somewhat doubtful whether we should talk any more of separate individuals!' "

" 'You're home, John!' "[25]

Bandwidth alone does not imply intelligence. Television uses a large bandwidth (about six megahertz per channel), but there is not much intelligence at either end. On the other hand, all known systems that exhibit intelligent behavior rely on the communication of information—two people discussing a problem, exchanging less than 100 bits per second, or the unfathomable number of bits per second exchanged among the 100 billion neurons within an individual brain. Information does not imply intelligence, and communication does not imply consciousness. The implications go the other way.

Irving J. Good, colleague of Alan Turing at both Manchester and Bletchley Park, although shying away from a theory of consciousness, which he considered, in the words of Turing, to be more a matter of "faith" than proof, addressed the relations between consciousness and communication in 1962: "The identification of consciousness with the operation of certain types of communication system has various consequences. We seem obliged to attribute more consciousness to two communication systems than to either separately, in fact it is natural to assume additivity. If the two systems, apart from their own operation, are in communication, then this will perhaps increase the amount of consciousness still further. The extra consciousness may reasonably be identified with the rate of transmission of information from one system to the other.... There may be a different kind of consciousness at each level, each metaphysical to the others ... the total consciousness may not have position in space, but it does have a sort of combinatorial topology, like a set of spheres with interconnecting tubes.... Thus, consciousness is a varying pattern of communication of great complexity [and] in order to decrease the complexity of the description of a communication network we may sum the flow of information over each given channel for a certain period of time."[26]

In his 1965 speculations on the development of ultraintelligent machines, Good, who recalled that "at the latest, by 1948, I had read both the books by Olaf Stapledon,"[27] saw wireless communication as the best way to construct an ultraparallel information-processing machine. "In order to achieve the requisite degree of ultraparallel working it might be useful for many of the elements of the machine to contain a very short-range microminiature radio transmitter and receiver. The range should be small compared with the dimensions of the whole machine. A 'connection' between two close artificial neurons could be made by having their transmitter and receiver on the same or close frequencies. The strength of the connection would be represented by the accuracy of the tuning. The receivers would need numerous filters so as to be capable of receiving on many different frequencies. 'Positive reinforcement' would correspond to improved tuning of these filters."[28]

Good was writing at the dawn of high-speed data communications, when most bits were still stored or transferred by punching holes in paper cards or tape, as had been done at Bletchley Park. The thicket of wires that had to be untangled every time the cryptanalysts wished to change the programming of the Colossus convinced Good from the beginning that wiring could take intelligence only so far. As computers coalesce into larger and faster networks, they are behaving more and more like the elemental units in Good's wireless ultraintelligent machine. We see the wires plugged into the wall and think of the architecture as constrained by the hardwired topology that the physical connection represents, whereas computationally, our machines belong to a diffuse, untethered cloud of the kind that Good envisioned as the basis of an ultraintelligent machine. All our networking protocols—packet switching, token ring, Ethernet, time-division multiplexing, asynchronous transfer mode, and so on—are simply a way of allowing hundreds of millions of individual processors to tune selectively to each others' signals, free of interference, as they wish.

Paul Baran, pioneer of packet switching, sees the relations between computers and communications advancing along similar, wireless lines. You can plug only so many things at one time into your wall. As everything from taxicabs to telephones to televisions to personal digital assistants becomes connected to the network, universal—and microminiature—wireless is the only way to disentangle the communications web. "But there's not enough wireless bandwidth to go around," say the skeptics, citing the billions of dollars raised whenever a few slivers of radio spectrum are auctioned off. Baran disagrees. "Tune a spectrum analyzer across a band of UHF frequencies and you encounter a few strong signals. Most of the band at any instant is primarily silence, or a background of weaker signals . . . much of the radio band is empty much of the time! The frequency shortage is caused by thinking solely in terms of dumb transmitters and dumb receivers. With today's smart electronics, even occupied frequencies could potentially be used."[29]

Baran made a similar argument in 1960, advising the government to build an all-digital, packet-switched data network instead of throwing good money after bad trying to blast-harden the centralized, circuit-switched network developed for analog transmission of voice. In both cases, he was right. Yet the regulatory establishment continues to treat radio spectrum as real estate to be sold to the highest bidder, instead of as an ocean across which low-powered, agile small craft can deliver information efficiently by following a few simple rules to keep out of each other's way. "The number of geographically dispersed users that can be simultaneously accommodated by a fixed spectrum varies as the inverse square of the transmission distance," Baran has pointed out, predicting that the meek and unlicensed may, in the end, inherit the earth. "Cut the range in half, and the number of users that can be supported is [quadrupled]. Cut the range by a factor of ten, and 100 times as many users can be served. . . . In other words, a mixture of terrestrial links plus shorter range radio links has the effect of increasing by orders of magnitude the usable frequency spectrum."[30]

While technologies such as low-earth-orbit satellite networks have received wider attention, Baran has been working to eliminate telecommunications bottlenecks in down-to-earth ways. The growth of a communications network, like any other arborescent, dendritic system, is driven by what happens at the root hairs, the leaf tips, the nerve endings, or, in telecommunications jargon, the network tails. The limit to network growth is peripheral; it is known as the last mile problem, or how to make the connection reaching the subscriber at the end.

A decade ago Baran saw two opportunities to grow new, or better utilize existing, tails. First was the cable-television network, which currently enters 63 percent of U.S. households and reaches the driveways of all but 7 percent. Coaxial television cable can transmit up to a gigahertz (one thousand megahertz) over short distances, enough for five hundred channels if digitally compressed. Most of this bandwidth is vacant most of the time. Baran founded Com21, Inc., in 1991 to develop the ultrafast packet switches and strategic alliances necessary to deliver a broadband digital communication spectrum over coaxial cable to the home, with a fiber-optic backbone linking the head ends of the local tails. Among the various schemes offering to provide broadband network growth, hybrid fiber-coaxial offers the path of least resistance because much of the infrastructure already is in place. What to do with all this bandwidth is a different problem, but history has shown that as bandwidth becomes available, the digital ecology swiftly takes root and grows.

Baran also founded (in 1985) a company named Metricom, better known by the name of its wireless network, Ricochet. A wireless, packet-switched, spread-spectrum digital communications network, Ricochet takes an extreme grassroots approach. It operates over small, collectively intelligent digital repeater-transceivers, about the size of lunch boxes, which perch inconspicuously on lampposts, drawing a trickle of power from the utility grid. The repeaters are situated about a quarter of a mile apart. Wherever large numbers of users congregate, you add more boxes here and there, and an occasional gateway to the local backbone of the net.

The whole system operates, unlicensed, within the 902–928 megahertz frequency band, under Federal Communications Commission rules that forbid the use of any given frequency for more than four hundred milliseconds at a time. The same adaptive-routing techniques that Baran first suggested in 1960, breaking a message up into packets that hop from mainframe to mainframe across a military communications net, are now used to convey messages not only by hopping from lunch box to lunch box, but from frequency to frequency (about ten times per second) at the same time. The twenty-six available megahertz is divided into 162 channels of 160 kilohertz each, and messages are divided into packets of 4,096 bits. From the point of view of an individual packet, not only is there a huge number of physically distinct paths from A to B through the mesh of lunch boxes, but there are 162 alternative channels leading to the nearest lunch box at any given time. The packet chooses a channel that happens to be quiet at that instant and jumps to the next lamppost at the speed of light. The multiplexing of communications across the available network topology is extended to the multiplexing of network topology across the available frequency spectrum. Communication becomes more efficient, fault tolerant, and secure.

The way the system works now (in a growing number of metropolitan areas—hence the name) is that you purchase or rent a small Ricochet modem, about the size of a large candy bar and transmitting at about two-thirds of a watt. Your modem establishes contact with the nearest pole-top lunch box or directly with any other modem of its species within range. Your computer sees the system as a standard modem connection or an Internet node, and the network, otherwise transparent to the users, keeps track of where all the users and all the lunch boxes are. This knowledge is fully distributed; there is no centralized intelligence or control. The modems and repeaters communicate at up to 100 kilobits per second; the user nets about 20 kilobits per second of usable communications, at no marginal cost above the flat monthly subscription, now about a dollar a day. The modems and repeaters cost several hundred dollars each. The system scales gracefully; the more users, the less expensive and more efficient it gets. It does not have to acquire expensive spectra in new markets as it grows.

Whether Metricom succeeds or fails, it demonstrates a communication system that uses less power, but becomes more powerful, as its nodes become larger in number but smaller in physical scale. Life as we know it is composed of small, cellular units not because larger units are impossible to build, but because smaller, semiautonomous units are faster, cheaper, easier to program, and easier to replace. Animals have become larger, but cells have stayed small. Everything in one of the lunch boxes could coalesce into a single component, a microcommunicator that, if produced in similarly large quantities as microprocessors, might cost as much to package as to make. A telepathic, subvital unit. All the economies that Baran demonstrated by adaptive message-block switching over microwave links between defense command centers are applicable to exchanging data directly between microprocessors. The ultimate decentralization of network architecture is for every processor to become a node.

Semiconductor technology originated nearly a century ago with the "cat's-whisker" crystal detector, which allowed the reception of coded signals via radio-frequency disturbances in the electromagnetic field. A wireless web of microprocessors is now relentlessly taking form. The "last centimeter" problem is keeping us a healthy distance from telepathy among human beings, but we are fast approaching telepathy among machines. Every last bit of isolated intelligence—

your living-room thermostat, your gas meter, your stereo system, the traffic light down the street—has an affinity for other bodies of intelligence, a faint attraction to all sources of information within reach.

Most of the time, despite the perception of being inundated with information, we will remain out of the loop. Human beings have only limited time and ability to communicate: you can watch television, check your E-mail, and talk on your cellular phone at the same time, but that's the limit. We are now the bottleneck—able to absorb a limited amount of information while producing even less, from the point of view of machines. While a person types full speed, the average microprocessor goes through millions of cycles between one keystroke and the next. This difference is what keeps humans human, and machines machines. "If we were ever to put all our brains together in fact, to make a common mind the way the ants do," warned Lewis Thomas, "it would be an unthinkable thought, way above our heads."[31]

Distributed intelligence, or composite mind, is a nebulous idea. On the other hand, we do not know of any intelligence that is *not* distributed, or any mind that is *not* composite. Carrying these concepts back toward the dawn of time, Olaf Stapledon left an unpublished, early draft of *Star Maker,* in which he speculated on the development of mind among the first, nebulous pregalactic structures in the universe, an exercise in imagination that might help us in understanding the nebulous predicament of any ultraintelligent machines—wired or wireless—that might be developing here on earth.

"To understand the mentality of the nebulae," wrote Stapledon, "one must bear in mind three facts which make them differ through and through from human beings. They do not succeed one another in generations; they are not constrained by economic necessity; the great majority of them have reached maturity in ignorance of other minds. ... With the nebulae there is no distinction between the growth of individuals and the evolution of the race. The life and memory of each nebula reaches back to the racial dawn. . . . The nebulae are in a sense 'nearer to God' than any man can ever be."[32]

Microprocessors, like neurons—quiet in death, or coma, but not in sleep—stay constantly on the alert. Unlike us, they have unlimited memory, and unlimited time, to fill. Hard-disk storage now costs less than ten cents a megabyte, and as fast as things are being stored, at the risk of being buried forever, more efficient methods are being evolved, so that they can be retrieved. Does this herald, as H. G. Wells predicted, a global consciousness? Maybe, or maybe not, but one thing is certain—global unconsciousness comes first.

"So perfectly organized was the life of the minded swarm that all routine activities of industry and agriculture had become, from the point of view of the swarm's mind, unconscious, like the digestive processes of a human being," wrote Olaf Stapledon, describing the minute, electromagnetically linked creatures inhabiting one of *Star Maker*'s plurality of worlds: a massive planet where extreme gravity—and the absence of oceans—prevented the development of creatures able to support large brains. "The little insectoid units themselves carried on these operations consciously, though without understanding their significance; but the mind of the swarm had lost the power of attending to them. Its concern was almost wholly with such activities as called for unified conscious control."[33]

Until we understand our own consciousness, there is no way to agree on what, if anything, constitutes consciousness among machines. The subject leads us into nonfalsifiable hypotheses, where the scientific method comes to an end. Three results are possible, given any supposedly conscious machine. Either the machine says, "Yes, I am conscious," or it says, "No, I am not conscious," or it says nothing at all. Which are we to believe? All we can do at this point is use our imaginations. And in this Olaf Stapledon was sixty years ahead. "Within the minded group, the insectoid units were forever dying off and giving place to fresh units," Stapledon explained, "but the mind of the group was potentially immortal."[34] This is true of all large, enduring communications networks—whether the nodes are as inscrutable as neurons, as intelligent as human beings, or as dumb as microprocessors that sit on lampposts bouncing messages around and not doing much of anything else.

FIDDLING WHILE ROME BURNS

"Then there is electricity;—the demon, the angel, the mighty physical power, the all-pervading intelligence!" exclaimed Clifford. "Is that a humbug too? Is it a fact—or have I dreamed it—that, by means of electricity, the world of matter has become a great nerve, vibrating thousands of miles in a breathless point of time? Rather, the round globe is a vast head, a brain, instinct with intelligence! Or, shall we say, it is itself a thought, nothing but thought, and no longer the substance which we deemed it?"

—NATHANIEL HAWTHORNE[1]

Since the dawn of technology humans have endowed artifacts with mind. In our collective imagination, inhabited by objects, animals, and now machines, mind has rarely been held as an exclusively human preserve. Mind has prevailed until recently as a quality distributed among all things, captured one lifetime at a time and then returned. Human language and memory have extended our possession of this instant so that most of us now live deeply embedded within the extended moment that our consciousness and culture represent. Only our designated prophets bring back something from the edge. "About ourselves there always lingers a penumbral rainbow ... which can be dissected from no single brain," wrote Loren Eiseley in 1970. "Something, the rainbow dancing before his eyes, the word uttered by the cave fire at evening, eludes us and runs onward. It is gone when we come with our spades upon the cold ashes of the campfire four hundred thousand years removed."[2]

In our various mythologies we have toyed with the prospect of mechanical intelligence, as childhood playthings anticipate the use of tools. In medieval times an Arabic fable, drifting through the centuries, became attached to the pope Silvester II, a man of great mathematical and mechanical abilities who died in the year 1003. Silvester, or Gerbert as he was known before being elevated to pope in the year 999, helped to introduce Arabic numerals and arithmetic into Europe, and, as reported by William of Malmesbury in the twelfth century, "he gave rules which are scarcely understood even by laborious computers."[3] He led the famous school at Reims, helped to secure tenure for research among the universities, and constructed mathematical instruments, a steam-driven organ, and mechanical clocks. He was rumored, sometimes darkly, to have invoked intelligence among things not born but built. According to William of Malmesbury, Silvester constructed a speaking head, which "spake not unless spoken to, but then pronounced the truth, either in the affirmative or the negative."[4] Delivering but one bit of information at a time, this oracle communicated with utmost economy and was always right.

In the thirteenth century this fable descended to Roger Bacon (ca. 1214–1292), an English scholar whose encyclopedic tastes reached beyond astrology and alchemy to embrace sciences far ahead of his time. Said to have been imprisoned for fifteen years by his own Franciscan brothers for the novelty of his ideas, he became known as Doctor Mirabilis, though without a shred of evidence that he ever had anything to do with a speaking mechanical head. But the legend stuck, imprinted first by the anonymous *Famous History of Frier Bacon* and deepened in the sixteenth century by *Friar Bacon and Friar Bungay* (1594), a play by Robert Greene. As the story goes, Bacon undertook to preserve England against conquest, and provide himself with everlasting fame, by constructing a wall of brass about the entire country, which, his studies indicated, could be achieved by enlisting the intelligence of a brazen head.

"To this purpose he got one Frier Bungey to assist him," it is explained, "who was a great scholar and a Magician, (but not to compare to Frier Bacon) these two with great study and pains so framed a head of Brass, that in the inward parts thereof there was all things like as in a natural man's head: this being done, they were as far from perfection of the work as they were before, for they knew not how to give those parts that they had made, motion, without which it was impossible that it should speak: many books they read, but yet could not find out any hope of what they sought, that at the last they concluded to raise a spirit, and to know of him that which they could not attain to by their own studies."[5]

Repairing to a nearby wood, they raised a reluctant, uncooperative "Devil" who, under pain of certain unpleasantries, disclosed the required formula, but refused to specify the length of time it would take for the process to take effect. "If they heard it not before it had done speaking," they were warned, "all their labour should be lost." Bacon and Bungey followed the devil's instructions exactly and waited three weeks, with no results. Then Bacon assigned his servant Miles to keep a close watch on the brass head so the two magicians could take a nap.

Miles amused himself with a pipe, song, and drum while his master slept, and then "at last, after some noise the head spake these two words, TIME IS. Miles hearing it to speak no more, thought his Master would be angry if he waked him for that, and therefore he let them both sleep, and began to mock the head in this manner: Thou Brazen-faced head, hath my Master took all this pains about thee, and now dost thou requite him with two words, TIME IS: had he watched with a lawyer as long as he hath watched with thee, he would have given him more, and better words then thou hast yet, if thou canst speak no wiser, they shall sleep till dooms day for me."[6]

Miles kept mocking the brass head: "Do you tell us Copper-nose, when TIME IS? I hope we Scholars know our times, when to drink, when to kiss our hostess, when to go on her score, and when to pay it, that time comes seldom," and so on. After half an hour of this "the head did speak again, two words, which were these: TIME WAS." Miles still would not wake his master, saying, "if you speak no wiser no Master shall be waked of me," and acted like a fool for another half hour. Then, without warning, all hell broke loose: "This Brazen Head spake again these words; TIME IS PAST: and therewith fell down, and presently followed a terrible noise, with strange flashes of fire, so that Miles was half dead with fear: at this noise the two Friers awaked, and wondered to see the whole room so full of smoke, but that being vanished they might perceive the brazen head broken and lying on the ground."[7] Thus Bacon's great project came to an end.

The story of Friar Bacon and the brazen head, however apocryphal, remains a fable for our time. Since the dawn of computers scientists have raised one species of spirit after another, seeking to have the mystery of intelligence revealed. When hopes are up, our Bacons and Bungeys have gathered openly, performed their incantations, and then retired, leaving their attendants to keep watch (and bear the costs). Thirty-five years ago neurologist Warren S. McCulloch, the architect with Walter Pitts of the first rigorous theory of neural nets, delivered ten pronouncements derived from twenty years of searching for "Where Is Fancy Bred." McCulloch's tenth commandment: "We will be there when the brass head speaks."[8]

Cloaked in symbols as arcane as those of the alchemists, Bacon and Bungey's successors continue to decipher their forefathers' instructions, keeping the foundries running day and night. Crystals are drawn from the crucible by workers masked and gowned against the risk of bringing the imperfection of our world to the kingdom of machines; diamond saw blades slice these crystals into wafers on which spells are cast by ultraviolet light. Some of the sorcerers work in silicon and some work purely in code, but when the two halves of this magic are brought together, still the brass head refuses to speak.

The skeptics have it that we are no closer than Bacon and Bungey to achieving the transmutation of metal into mind. Optimists believe it is just a matter of enough time, enough logic, the right coding, or some contagious spark of wisdom we haven't put our finger on yet. Others believe that we are playing the role of Miles, mocking the long-awaited signs while our master remains asleep.

In the 1950s computers demonstrated their dexterity at manipulating very large numbers over minute increments of time. "Time is!" they seemed to say—but after giving the matter some thought, we decided not to awake our master for mere arithmetic, just yet. Twenty years passed. In the 1970s computers began to reproduce themselves in automated factories in accordance with von Neumann's principles of how automata can grow more complicated from one generation to the next. "Time was!" their advancing numbers proclaimed—but we decided that mere spreadsheets and word processors did not merit raising the alarm. Another twenty years passed. Computers, now teeming like herring in early spring, began pooling their intelligence, exchanging states of mind in the blink of an eye, half a dozen languages removed from those that we can comprehend. Only an esoteric fraternity, uttering one line of code at a time, still holds congress with the machines. "Time is past!" can be read between the lines. But the warning goes unheeded as we stand transfixed, like monkeys given a mirror, by the novelty of our own image reflected in the surface of the web. When the smoke clears and the master wakes, the computer as disembodied head will have disappeared, replaced by a diffuse tissue enveloping us in nebulous bits of meaning, as neurons are enveloped in electrolyte by the brain.

There are two approaches to embodying intelligence, whether as a brain or as a brazen head. The alternatives correspond to two different approaches to building a boat. To build a kayak, you assemble a skeleton and then give it a skin that allows it to float, just as the architectural framework of a computer is fitted, by evolution or by design, with an envelope of code. To build a dugout, you grow a tree and then remove everything, one chip at a time, except the boat. This is how nature creates her intelligences, by spawning an overwhelming surplus of neurons and then selectively pruning them to leave a network that, if all goes well, becomes a mind. As computers are replicated by the millions, they are aggregating into structures whose design bears nature's signature in addition to our own.

Within this computational matrix, whether viewed as code subsisting on processors or processors subsisting on code, organization is arrived at as much by chance as by design. Most of the connections make no sense, and few make any money except in circuitous ways. Critics say that the World Wide Web is a passing stage (right) and doomed to failure because it is so pervaded by junk (wrong). Yes, most links will be relegated to oblivion, but this wasteful process, like many of nature's profligacies, will leave behind a structure that could not otherwise have taken form. "I could not discover the mechanism of this steady internal evolution," wrote Stapledon in pondering the mentality of the nebulae. "But one point seemed to me certain. Natural selection played an important part within each nebula, favoring some experiments in vital organization and destroying others."[9] The World Wide Web, a primitive metabolism nourished by the substance of the Internet, will be succeeded by higher forms of organization feeding upon the substance of the World Wide Web.

The prevailing approach to artificial biology views computers as terraria in which digital creatures are evolved. This assumption fits conveniently within the laboratory, but as a model of the digital universe it provides only a silhouette. Robert Davidge, of the University of Sussex, reversed the perspective in a paper titled "Processors as Organisms," published in 1992. He suggested that we "shift our view from the programs formed in the memory to the processor itself. . . . Consider the actual processor to be the organism and the memory of instructions to be its environment for exploration. . . . The static computer processor requesting instructions down the memory bus can become an organism moving through the memory. It is the same procedure, but it entirely changes the way we regard the results."[10] Both perspectives are essential to understanding the origins of artificial life.

"If we choose to think in the frame of mind of a biologist then we can begin to see biological types of process in the artificial creations that surround us," explained Davidge. But a computer's movement through its memory is a one-dimensional process. "To be of any interest biologically," Davidge noted, "the memory must be 2-dimensional not 1-dimensional as in all standard stored-program

computers. . . . If we get rid of the idea that the processor exists to
execute our program then we can let it move in a 2-D or 3-D space of
instructions and the motional behaviour will become a continuous
track through space."[11] Shortly after Davidge published his speculations this two- or three-dimensional memory and instruction space
was realized, suddenly, in the form of the World Wide Web. The Web
allows code to move freely through the visible universe of processors,
and it allows processors to move freely through the visible universe of
code. The result is more than the sum of the parts. "Life," as Samuel
Butler observed in 1887, "is two and two making five."[12]

Coding and processing, like matter and energy in conventional
physics, are related manifestations of an underlying field. This computational field is observed and measured in bits. A bit is the
fundamental unit of information—the difference between two dis-
cernible alternatives, perceived as change or choice. The computational universe and the universe of time and space in which we live
intersect by means of two kinds of bits: bits that represent a difference
between two things at the same time; and bits that represent a
difference between one thing at two different times.

The power of computers—whether the strictly regulated machi-
nations of a Turing machine or the amorphous intelligence residing in
our heads—derives from their ability to form maps between sequence, arranged in time, and structure, arranged in space. Memory
and recall, no matter what their form, are translations between these
two species of bits. "Memory locations," according to Danny Hillis,
"are just wires turned sideways in time."[13] In this correspondence
between sequence and structure lies the basis not only of computation
and memory, but also of organic life, based on the translation of
sequences (nucleotides) into structures (proteins), with natural selection the mechanism for debugging the source code and translating
improvements back from the structure of the organism to the sequence of its genes. Computers are speeding the process up.

In Alan Turing's minimal example, translation between structure
and sequence is executed one bit at a time. The Turing machine scans
one square on its tape, reads one bit of information, makes a corresponding change in its state of mind, and, in accordance with its instructions, writes or erases one bit of information on its tape. When the
next moment arrives, it goes through this process again. The Turing
machine and its visible universe cross paths one symbol at a time.

As bandwidth measures the capacity to communicate information
from one place to another, so it is possible to assign a magnitude to
the amount of information that a Turing machine, or other organism,

is able to scan as it moves from one moment to the next. Extending Turing's terminology, we may call this the machine's depth of mental field. Going one step further, it is possible to quantify how much information can be scanned at a single moment, multiplied by the number of consecutive moments that can be comprehended within the machine's state of mind. This scale of mental capacity allows us to make comparisons between minds that may be operating at completely alien velocities in time.

In the four-dimensional universe of space and time, we are confined to a three-dimensional surface. Only one moment exists in our reality; the existence of other moments is evident only through the constructs of our mind. We breathe one lifetime at a time within a thin atmosphere condensed out of the surrounding sequence of events. All our devices for translating between sequence and structure serve to extend this atmospheric depth—the history of life on earth is compressed within sequential strands of DNA; culture is accumulated in the form of language; somehow our brains preserve the sequence of our lives from one moment to the next. As far as we know mind and intelligence exist on an open-ended scale. Perhaps mind is a lucky accident that exists only at our particular depth of field, like some alpine flower that blooms between ten thousand and twelve thousand feet. Or perhaps there is mind at elevations both above and below our own.

"It may be that the cells of which we are built up . . . each one of them with a life and memory of its own . . . reckon time in a manner inconceivable to us," suggested Samuel Butler in 1877. "If, in like manner, we were to allow our imagination to conceive the existence of a being as much in need of a microscope for our time and affairs as we for those of our own component cells, the years would be to such a being but as the winkings or the twinklings of an eye." Writing in *Life and Habit* on the eve of his estrangement from Charles Darwin, Butler sought to encompass Darwinism—from the life of germs to the germination of species—within a framework of all-pervasive mind. "What I wish is, to make the same sort of step in an upward direction which has already been taken in a downward one, and to show reason for thinking that we are only component atoms of a single compound creature, LIFE, which has probably a distinct conception of its own personality though none whatever of ours."[14]

Olaf Stapledon believed that the mind of the individual and the mind of the species need not remain estranged. "Our experience was enlarged not only spatially but temporally in a very strange manner," explains the narrator of *Last and First Men*, concerning the composite mind toward which our species had inexorably evolved. "In respect of temporal perception, of course, minds may differ in two ways, in the length of the span which they can comprehend as 'now,' and in the minuteness of the successive events which they can discriminate within the 'now.' As individuals we can hold within one 'now' a duration equal to the old terrestrial day; and within that duration, we can if we will, discriminate rapid pulsations such as commonly we hear together as a high musical tone. As the race-mind we perceived as 'now' the whole period since the birth of the oldest living individuals, and the whole past of the species appeared as personal memory, stretching back into the mist of infancy. Yet we could, if we willed, discriminate within the 'now' one light-vibration from the next."[15]

Both Samuel Butler and Olaf Stapledon saw that mind, once given a taste of time, would never rest until eternity lay within its grasp. Thus we pursue those relations between sequence and structure that allow us to escape time's surface, venturing into that ocean that separates eternity from the instant in which we exist. Mathematics and music are two of the vehicles that assist us in our escape. Mathematics is available to a few; music is available to all. Mathematics allows us to assemble mental structures by which we comprehend entire sequences of logical implication at once. Music allows us to assemble temporal sequences into mental scaffolding that transcends the thinness of time in which we live. Through music, we are able to share four-dimensional structures we are otherwise only able to observe in slices, one moment at a time.

When computers began to multiply in the 1950s, artificial intelligence was believed to be just around the corner. Artificial music was surely almost as close at hand. Forty years later, artificial intelligence is still ahead. Electronics has enlarged our repertoire of instruments but has failed to produce anything more than a sympathetic resonance with the musical nature of our minds. Artificial music, so far, has the same relation to real music that animation has to life. Computers have perfect pitch, perfect timing, and perfect memory—achievements only the best musicians attain—yet something about music leaves them cold. The gap between the natural language of human beings and the higher-level languages and formalizations used by machines is slowly being bridged. But our music remains a foreign language to our machines.

How did music first evolve, and how might it develop further or evolve anew in different form? Neurophysiologist William Calvin has suggested that music is a by-product of the need to store complex sequences of motor instructions in serial buffers in the brain. "Movement command buffers, essential for planning ballistic movements (so fast that sensory feedback arrives too late to effect corrections), were surely under selection for throwing," Calvin explained. "For organisms that need to be both large (meters of conduction distance) and fast, one often needs the neural equivalent of an old-fashioned roll for a player-piano. We carefully plan during 'get-set' to order to act without feedback. If those buffers are capable of sequencing other things when not needed for throwing-hammering-clubbing-kicking, then one might expect augmentation of such sequential abilities as stringing words together into sentences, or concepts into scenarios."[16] Calvin suggested that parallel command buffers—he used the analogy of trains being assembled on the parallel tracks of a freight yard—may have stored alternative sequences at the same time. Fitter or more attractive sequences would be selected to survive, reproduce, and recombine.

Nature loves dual, or multiple, functions. Once motor-control sequences were stored in the same buffers as sequences of sound—a plausible scenario given the layout of mammalian brains—a cascade of developments was under way. The slow, evolutionary arms race between physically distinct individuals was accelerated by evolutionary processes acting within individual brains. To hit a target at even a modest distance, a thrown weapon must be released at exactly the right moment; the required submillisecond timing can be achieved only through parallel arrays of neurons that collectively smooth out the temporal jitter characteristic of a few neurons operating on their own. Precise relations between timing and frequency—the raw material of music—would be facilitated by the huge, parallel buffers of the expanding human brain, which, as we know but have yet to explain, tripled in volume over a short span of evolutionary time.

Larger brains allowed abstract four-dimensional processes to take root and grow. Individuals better able to plan, compare, rehearse, remember, and execute complex sequences of movements were better able to survive, and so these abilities were reinforced. But neither mind nor music, as far as we know, can be sparked in one brain alone. On one level, music may have evolved as a way to exercise and communicate these abilities; on another level, music may have evolved because these mental structures, able to reproduce themselves across time and distance, developed a life of their own.

To me, Calvin's hypothesis strikes a chord. My grandfather George Dyson (1883–1964) was a professional musician who first secured fame and lasting fortune by his mastery of the art of throwing bombs. He wrote dozens of well-received works of music and three successful books, but none of them ever matched the sales of his instructions for hurling a kilogram of iron and high explosive thirty meters through the air. When war broke out in August 1914, my grandfather, unlike Olaf Stapledon or Lewis Richardson, had no second thoughts—he immediately signed up. He was commissioned as a lieutenant in the Ninety-ninth Infantry Brigade of the Royal Fusiliers, stationed at Tidworth on Salisbury Plain, and given the job of training infantry in the use of grenades before they went to France.

The hand grenade was viewed as an archaic, unsporting weapon, of dubious utility on the modern battlefield now that rifles and machine guns could kill precisely at distances far greater than the range of a grenade. But a season or two of trench warfare was about to change all that. Dyson was in the right place at the right time. "A weapon has had to be evolved that shall have as much destructive force as possible, combined with such a high angle of descent as will render the mere depth of a trench no efficient protection," he wrote in 1917.[17] Unable to locate either experts or grenades, Dyson became an expert and designed his own reusable practice grenades, heavy iron castings loaded with just the right amount of gunpowder and clay. As the son of a Yorkshire blacksmith, he knew how to approach local suppliers without letting formalities get in the way. He constructed a training ground complete with fire trenches, traverses, listening posts, and machine-gun emplacements and set about developing the talents of his soldiers with all the enthusiasm he had devoted to his best music students before the war.

"A very high standard of accuracy is required," he wrote in 1915. "The difficulties are greatly augmented when, as is usually the case, throwing has to be done under complete cover, and according to directions given verbally by an observer. It is impossible to exaggerate the importance of throwing practice of every kind. . . . Nothing can excuse inaccurate throwing, and instructors must not be satisfied until the thrower can, from behind cover, and in obedience to the command of an observer, throw missiles of varying weight and size into a specified trench at any reasonable distance or in any direction."[18]

Dyson had been born with perfect pitch for music; he now did his best to cultivate perfect pitch in the throwing of grenades. Timing was everything. Although trenches were stationary targets, the unforgiving nature of a four-second fuse added some of the temporal challenges faced by a hunter throwing a projectile at moving prey. "When using live bombs the thrower will hold the bomb in his hand with the arm extended. The carrier who will always be with him, will light the fuze and as soon as it is lit will tap the thrower's arm as a signal that the bomb is ready to throw, and the thrower will time his throw so that the bomb falls into the objective just before it explodes."[19] Dyson held competitions between teams of grenadiers, organized like cricket matches but producing a lot more noise. "Instructors must take the greatest care to ensure that familiarity with explosives does not lead to carelessness in handling them," he warned. "Grenadiers must be constantly reminded of the great danger in swinging percussion grenades. A hit or graze on some part of the trench may easily be fatal to the thrower. No preliminary swinging will be allowed."[20] To throw accurately without a windup was like performing a difficult piece of music without a rehearsal.

Dyson's methods attracted the attention of his superiors and were duplicated far and wide. His notes were issued in 1915 as a small pamphlet, *Grenade Warfare: Notes on the Training and Organization of Grenadiers*, selling for sixpence and designed to be tucked into the pockets of soldiers heading to the front. An expanded edition, costing fifty cents, was published in New York in 1917. As usual in military training manuals, only passing mention is given to what happens after the bombs explode. An appended section on bayonet fighting notes that "when the thrower has thrown his bomb into the objective trench, the bayonet men must be ready to take instant advantage of the temporary demoralization of the enemy caused by the explosion and clear the way for a similar attack on the next section of trench."[21]

Royalties accumulated a fraction of a sixpence at a time. Dyson survived a tour of duty on the front lines in France, returning shell-shocked but otherwise unscathed. By the time of the armistice in November 1918 an entire generation had been maimed by a war in which, as Garet Garrett put it, "God was on the side of the most machines."[22] Taking the side of the foot soldier, my grandfather demonstrated that throwing things by hand still had a place in the age of airplanes and tanks. He returned to civilian life comfortably well off, raised a family, and, as director of the Royal College of Music during World War II, slept in a storeroom while leading the effort to maintain musical performances in London amid the blackout and the bombs. "Apart from the incendiary fire in the opera wing which, thanks to [Fred] Devenish and the fire-watchers, did not reach the main building, and the later blitz which smashed two hundred windows, we had no serious damage," he reported. "But we were on the edge of destruction for months on end."[23] When he published his autobiography in 1954 he titled it *Fiddling While Rome Burns*. The creativity of music and the destruction of war are opposing expressions of the human spirit, yet my grandfather found room for both.

Dyson, who wrote, "I cannot remember the time when I could not read music and hear what I had read," began playing the organ professionally at age thirteen and worked among the great cathedrals all his life. But "he was not particularly religious," my father recalled. "He always said that music was as close to religion as he could get." He knew that music transcended everyday existence, but he never understood how, or why. These abilities remain unexplained. "The number of sounds and patterns that a sensitive musical brain can hold and recall at will is beyond either computation or understanding," Dyson wrote, citing abilities far exceeding his own. "There have been and are still conductors who by defect of sight cannot read a score while they are conducting it. Toscanini is one of these, and he comes to the rehearsal, even of a whole opera, with every detail of every part already registered in his mind."[24] Noting that it was physically impossible to play consistently in tune on early instruments, he commented that "Beethoven would never have heard, even if his own hearing had been perfect, even a tolerable performance, by our standards, of the greater works of his maturity. He, like his predecessors, wrote from his imagination. The music was created in his mind, and nowhere else."[25]

My grandfather regarded the origins of music and musical abilities as an impenetrable mystery, their purpose not to be explained in terms of anything else. "Why have we these powers of intuition and expression in sound which so completely transcend the normal use of our senses, and which appear to have neither a boundary or a meaning that can be rationally defined?" he asked. "The only theory of art which makes sense is that which acknowledges the specific creation of a new world, a world differing from, and not to be explained in terms of, any other worlds, material or physical, except that of the art in question. . . . The arts continue to develop themselves, and enlarge our faculties of perception and response, in their own right and on their own terms. The saint both pursues and creates religion. The scientist both seeks and makes truth. The artist evolves his own sense of order and expresses it by his craft. These are all new worlds, living their own lives according to their own laws. We cannot explain a world created by our imagination. It may have no material counterpart in life at all."[26]

The mystery of music brings us to the closing fable of this book. It is the work of Danny Hillis, architect of the massively parallel Connection Machine, and one of the closer approximations to a Doctor Mirabilis alive today. Fables survive in association with enduring questions. The question here is why the mystery of music runs even deeper than the mystery of mind. In a 1988 essay titled "Intelligence as an Emergent Behavior; or, the Songs of Eden," Hillis offered a parable in which, to simplify an already simplified story, music leads to the emergence of mind rather than the other way around.

"Once upon a time, about two and a half million years ago, there lived a race of apes that walked upright. In terms of intellect and habit they were similar to modern chimpanzees. The young apes, like many young apes today, had a tendency to mimic the actions of others. In particular, they had a tendency to imitate sounds. . . . Some sequences of sounds were more likely to be repeated than others. I will call these 'songs.'

"For the moment let us ignore the evolution of the apes and consider the evolution of the songs. Since the songs were replicated by the apes, and since they sometimes died away and were occasionally combined with others, we may consider them, very loosely, a form of life. They survived, bred, competed with one another, and evolved according to their own criterion of fitness. If a song contained a particularly catchy phrase that caused it to be repeated often, then that phrase was likely to be repeated and incorporated into other songs. Only songs that had a strong tendency to be repeated survived.

"The survival of the song was only indirectly related to the survival of the apes. It was more directly affected by the survival of other songs. Since the apes were a limited resource, the songs had to compete with one another for a chance to be sung. One successful strategy for competition was for a song to specialize; that is, for it to find a particular niche where it would be likely to be repeated.

"Up to this point the songs were not of any particular value to the apes. In a biological sense they were parasites, taking advantage of the apes' tendency to imitate. Once the songs began to specialize, however, it became advantageous for an ape to pay attention to the songs of others and to differentiate between them. By listening to songs, a clever ape could gain useful information. For example, an ape could infer that another ape had found food, or that it was likely to attack. Once the apes began to take advantage of the songs, a mutually beneficial symbiosis developed. Songs enhanced their survival by conveying useful information. Apes enhanced their survival by improving their capacity to remember, replicate, and understand songs. The blind forces of evolution created a partnership between the songs and the apes that thrived on the basis of mutual self-interest. Eventually this partnership evolved into one of the world's most successful symbionts: us."[27]

This explanation of how symbiosis between sequence (songs) and structure (apes) led to the evolution of mind is subject, like all good fables, to a wide range of interpretations. It can be interpreted as an explanation of how sequence (genes) and structure (metabolism) led to the development of organic life, or of how sequence (coding) and structure (computers) is leading to developments we can only begin to grasp. Computer software developed hierarchically, just as stored motor-control sequences may have led to increasing levels of abstraction in our brains. Fifty years ago the only electronic software in existence comprised a few strings of instructions occupying the storage buffers of the ENIAC, designed to calculate the launch window for firing artillery or dropping bombs. As Hillis's songs of Eden evolved from music into mind, digital coding has come of age, living its own lives according to its own laws. We have absolutely no idea what species of music or what species of mind will be the result. It is unlikely that it will be perceivable as music or as mind to us.

For a long, long time—much longer than we have been waiting for the brass head to speak—we have awaited the appearance of a higher intelligence from above or a larger intelligence from without. Arthur C. Clarke, Olaf Stapledon's distinguished successor, wrote his novel *Childhood's End* (1953) around the premise that an advanced intelligence would descend from outer space, bringing our childhood to an end. The Overlords, as they were called, absconded with the minds of our children, and humanity as we know it came to an end. We should heed this warning as we proceed to construct what amounts to an overmind. Alien beings are unlikely to bear any resemblance, mental or physical, to human beings, and it is presumptuous to assume that artificial intelligence will operate on a level, or a time scale, that we are able to comprehend. As we merge toward collective intelligence, our own language and intelligence may be relegated to a subsidiary role or left behind. When the brass head speaks, there is no guarantee that it will speak in a language that we can understand.

The evolution of languages is a central mechanism by which life and intelligence unfold. Over the past fifty years the digital universe has spawned a host of languages, the most successful being slightly better adapted not only to the workings of machines but to the human mind. Translation remains tedious and slow. Human beings are unlikely ever to speak the language of digital computers much faster than they did at the beginning, one line of code at a time. Indeed, when it comes to binary coding by human beings, it would be difficult to improve on the abilities of telegraph operators, who had a century's head start.

Languages are maps. By conveying information across distance and over time, or from one form of expression to another, languages derive the sustenance on which they feed and grow. Morse code provides a mapping between the alphabet and short, dot-dash strings; the genetic code translates between nucleotides and proteins; natural language translates between words and ideas; HyperText Markup Language (HTML) maps the topology of the Internet into strings of communicable code. Languages survive by hosting the reproduction of structures (letters, words, enzymes, ideas, books, or cultures) that in turn constitute a system sustaining the language from which they sprang. Music translates, in ways we little understand, between sequences of sounds and mental structures that have proved astonishingly successful at propagating themselves. That music does not map to any other known language does not render its meaning any less exact. "Music," answered Mendelssohn when asked about the meaning of his *Songs Without Words*, "is not thought too *indefinite* to be put into words, but, on the contrary, too *definite*."[28]

Von Neumann believed that the fundamental language of the brain is encoded by higher-order statistical relationships between trains of frequency-modulated pulses—just as the elements of music are encoded by the difference between two notes at the same time, or between one note and the next. In the evolution of mind, something like music came first, emerging from a primordial jungle of patterns vying to be exchanged and reproduced. Only later was this wilderness mapped by language into our ordered system of symbols and ideas. Mendelssohn can speak for all of us without words.

Hillis's metaphor of songs and apes can be extended by analogy to the digital universe without implying an artificial music that will sound like music to us. The song of the machine may be inaudible to our ears, invisible to our eyes, and unthinkable to our minds. Nonetheless, a more transparent language may grow in ways impossible to languages that have to be translated many times over in passing between the world of bits and the world of brains. Most evolutionary breakthroughs are the result of adapting abilities developed for something else. One means to bridge the communications disparity between computers and human beings may be to appeal directly to the pulse-frequency coding used within our brains. Instead of adding new layers of language in the attempt to facilitate communication, the breakthrough may come by peeling away the barriers of language to reveal what lies underneath. Music is the layer through which these foundations might be exposed. "A man may frame a Language, consisting only of Tunes and such inarticulate sounds, as no Letters can expresse," wrote John Wilkins in 1641. "If these inarticulate sounds be contrived for the expression, not of words and letters, but of things and notions, (as was before explained, concerning the Universall Character) then might there bee such a generall Language, as should be equally speakable, by all peoples and Nations; and so we might be restored from the second generall curse, which is yet manifested, not only in the confusion of writing, but also of speech."[29]

Children, not adults, bring new languages to life. Children learn to speak, read, and write before they learn to type, after which they are able to communicate, at the speed of a nineteenth-century teletypewriter, with one-hundred-megahertz machines. The speed with which young children learn to recognize new language or communicate in sign language shows how artificial a barrier this is. A dangerous imbalance exists, tempting an evolutionary jump that might be too far or too fast for our own good. Human–computer communication made its greatest advance in recent history via the mouse—allowing the user to do nothing more than point and click. By means of one constricted channel of communication, the mouse opened windows by which the entire computational ecology was transformed. For good or evil, natural selection now favors machines better able to communicate with children, and children better able to communicate with machines. The desktop light-guns from which today's mice descended first appeared in large numbers among the SAGE system users forty years ago. The pace has quickened ever since.

If all goes well, our children will be linked ever more closely to the myriad ganglia embedded in their lives, while remaining members of the human race. In the distant future, they may look back on us as children and wonder how, before symbiosis with telepathic machines, it was possible for us to communicate, or even think. But things could just as easily go the other way. "Evolution will take its course. And that course has generally been downwards," J. B. S. Haldane warned in 1928. "The majority of species have degenerated and become extinct, or, what is perhaps worse, gradually lost many of their functions. The ancestors of oysters and barnacles had heads. Snakes have lost their limbs and ostriches and penguins their power of flight. Man may just as easily lose his intelligence."[30] The choice is up to us. Erewhon is Nowhere, Samuel Butler warned. There is no turning back the clock.

Garet Garrett (1878–1954) published his pocket-sized but deeply cautionary book, *Ouroboros; or, the Mechanical Extension of Mankind,* in 1926. In the second chapter, titled "The Machine as If," he noted that "Either the machine has a meaning to life that we have not yet been able to interpret in a rational manner or it is itself a manifestation of life and therefore mysterious. We have seen it grow."[31] Garrett found it ominous that only half the world's population was still concerned with producing food: "The new, non-agricultural half is the industrial part; it is the part that serves machines."[32] Ouroboros was a mythical serpent that swallowed its own tail, an embodied contradiction that must, according to logic, either grow ever larger on its miraculous diet or, just as miraculously, consume itself and cease to exist.

Garrett saw machines as creations of the human mind, mindlessly unleashed. "The machine was not. He reached his mind into emptiness and seized it. Even yet he cannot realize what he has done. Out of the free elemental stuff of the universe, visible and invisible, some of it imponderable, such as lightning, he has invented a class of typhonic, mindless organisms, exempt from the will of nature."[33] Is the diffusion of intelligence among machines any more or less frightening? Would we rather share our world with mindless or minded machines? Garrett saw the growth of technology as an irredeemably selfish process, with "no hope of its being reformed ideally by mass intelligence." The only hope, in his analysis, lay in "a very curious suggestion that organisms now existing together in a state of permanent symbiotic union were once parasitic and learned better."[34] He concluded with a warning that deserves to guide us today: "In any light, man's further task is Jovian. That is to learn how best to live with these powerful creatures of his mind, how to give their fecundity a law and their functions a rhythm, how not to employ them in error against himself."[35]

Leviathan and Ouroboros were fellow mythological creatures: one whose power encompassed everything, the other that succeeded in swallowing itself. Technology has brought both Ouroboros and Leviathan to life. Is the diffuse mentality taking shape around us something new, or is it an ancient intelligence now awakened by speeding things up? Nature has always operated intelligently, but this intelligence has been perceived as either large and slow, as in evolution, or small and fast, as in quantum mechanics, leaving us alone on middle ground. "Does it not appear from Phaenomena that there is a Being incorporeal, living, intelligent, omnipresent," asked Isaac Newton at the conclusion of his *Opticks*, "who in infinite Space, as it were by his Sensory, sees the things themselves intimately, and throughly perceives them, and comprehends them wholly by their immediate presence to himself?"[36] Newton's God was incorporeal, as Hobbes's God was corporeal, and perhaps both these Gods are necessary, mirroring in their symbiosis the symbiosis between the songs and the apes.

Things have a way of surprising us. At the beginning of the twentieth century David Hilbert sought to establish the completeness of mathematics and instead precipitated the realization that the extent of mathematical truth can never be systematically confined. Evolutionists, who appeared to have dislodged mind and intelligence, are discovering that evolution is an intelligent process and intelligence an evolutionary process, rendering the separation less distinct. Technology, hailed as the means of bringing nature under the control of our intelligence, is enabling nature to exercise intelligence over us.

We have mapped, tamed, and dismembered the physical wilderness of our earth. But, at the same time, we have created a digital wilderness whose evolution may embody a collective wisdom greater than our own. No digital universe can ever be completely mapped. We have traded one jungle for another, and in this direction lies not fear but hope. For our destiny and our sanity as human beings depend on our ability to serve a nature whose intelligence we can glimpse all around us, but never quite comprehend.

Not in wilderness, but "in Wildness," wrote an often misquoted Henry David Thoreau, "is the preservation of the world."[37]

NOTES

Epigraph: Philip Morrison, "Interstellar Communication," in A. G. W. Cameron, ed., *Interstellar Communication: A Collection of Reprints and Original Contributions* (New York: W. A. Benjamin, 1963), 253.

ACKNOWLEDGMENTS

1. Freeman J. Dyson, *Origins of Life* (Cambridge: Cambridge University Press, 1985).

2. Verena Huber-Dyson, *Gödel's Theorems: A Workbook on Formalization* (Stuttgart: B. G. Teubner, 1991).

CHAPTER 1

1. Thomas Hobbes, *Leviathan; or, The Matter, Forme, and Power of a Commonwealth Ecclesiasticall and Civill* (London: Andrew Crooke, 1651), 1.

2. Ibid., 1.

3. Alexander Ross, epistle dedicatory to *Leviathan drawn out with a hook; or, Animadversions upon Mr. Hobbs, his Leviathan* (London: Richard Royston, 1653).

4. "The Judgment and Decree of the University of Oxford Past in their Convocation," 1683; in Samuel Mintz, *The Hunting of Leviathan* (Cambridge: Cambridge University Press, 1962), 61–62.

5. Hobbes, *Leviathan*, 1.

6. Ibid., 3.

7. Ibid., 371.

8. Ibid., 371–373.

9. Thomas Hobbes, in René Descartes, *Six Metaphysical Meditations; Wherein it is Proved that there is a God. And that Mans Mind is really distinct from his Body* (London: Benjamin Tooke, 1680), 119–120.

10. Ibid., 126–127.

11. Ross, epistle dedicatory to *Leviathan drawn out with a hook*.

12. Thomas Hobbes, as quoted in Isaac Disraeli, *Quarrels of Authors* (London: John Murray, 1814), 37.

13. Disraeli, *Quarrels of Authors*, 42.

14. Thomas Hobbes, 1662, *Considerations upon the Reputation, Loyalty, Manners, & Religion, of Thomas Hobbes of Malmsbury, written by himself, by way of Letter to a Learned Person* (London: William Crooke, 1680), 32.

15. Thomas Hobbes to Cosimo de' Medici, 6 August 1669, in Noel Malcolm, ed., *The Correspondence of Thomas Hobbes*, vol. 2 (Oxford: Oxford University Press, 1994), 711.

16. Samuel Pepys, 3 September 1668, *Diary and Correspondence of Samuel Pepys, F.R.S., Deciphered by Rev. J. Smith, A.M. from the original shorthand MS*, vol. 4 (Philadelphia: John D. Morris, 1890), 16.

17. John Aubrey, in *Aubrey's Brief Lives: Edited from the Original Manuscripts and with a Life of John Aubrey by Oliver Lawson Dick* (Ann Arbor: University of Michigan Press, 1949), 151.

18. Ibid., 156.

19. Steve Shapin and Simon Schaffer, *Leviathan and the Air-pump: Hobbes, Boyle, and the Experimental Life* (Princeton, N.J.: Princeton University Press, 1985), 344.

20. André-Marie Ampère, *Essai sur la philosophie des sciences, ou Exposition analytique d'une classification naturelle de toutes les connaissances humaines*, 2 vols. (Paris: Bachelier, 1834–1843).

21. Ibid., vol. 2, 141. (Author's translation).

22. Norbert Wiener, *Cybernetics; or, Control and Communication in the Animal and the Machine* (New York: John Wiley, 1948), 19.

23. Thomas Hobbes, *Elements of Philosophy: The first section, Concerning Body* (London: Andrew Crooke, 1656), 2–3.

24. Marvin Minsky, "Why People Think Computers Can't," *Technology Review* (November–December 1983): 64–70.

25. "Worldwide Semiconductor Unit Shipments," graph attributed to Integrated Circuit Engineering Corp., in *Standard & Poor's Industry Surveys: Electronics* (New York: Standard & Poor's Corp.), 3 August 1995, E25.

26. "Worldwide Demand for Silicon," graph attributed to Dataquest, Inc., in *Electronic Business Today* 22, no. 5 (May 1996): 39.

27. Linley Gwennap, "Revised Model Reduces Cost Estimates," *Microprocessor Report* 10, no. 4 (25 March 1996): 18, 23.

28. Price Waterhouse, Inc., *Technology Forecast: 1996* (Menlo Park, Calif.: Price Waterhouse Technology Centre, October 1995), 21.

29. "Worldwide DRAM Market in Billions of Units," graph attributed to Bernstein Research, Inc., in *Electronics* 68, no. 2 (23 January 1995): 4.

30. Donald Keck, "Fiber Optics: The Bridge to the Next Millenium," Corning Telecommunications *Guidelines* 10, no. 2 (Autumn 1996): 2.

31. U.S. Federal Communications Commission, *Fiber Deployment Update, end of 1995* (Washington, D.C., July 1996); *Fiber Optics,* an update to the update for 1996, U.S. Office of Telecommunications, March 1996.

32. Alex Mandl, talk given at the 1995 Platforms for Communication Forum, Phoenix, March 8, 1995.

33. W. Daniel Hillis, "Intelligence as an Emergent Behavior; or, The Songs of Eden," *Daedalus* (winter 1988) (*Proceedings of the American Academy of Arts and Sciences* 117, no. 1): 176.

34. H. G. Wells, *World Brain* (New York: Doubleday, 1938), xvi.

35. Ibid., 87.

36. Philip Morrison, "Entropy, Life, and Communication," in Cyril Ponnamperuma and A. G .W. Cameron, eds., *Interstellar Communication: Scientific Perspectives* (Boston: Houghton Mifflin, 1974), 180.

37. Irving J. Good, *Speculations on Perceptrons and other Automata,* IBM Research Lecture RC-115 (Yorktown Heights: IBM, 1959), 6. Based on a lecture sponsored by the Machine Organization Department, 17 December 1958.

38. Lynn Margulis and Dorion Sagan, *Microcosmos: Four Billion Years of Microbial Evolution* (New York: Simon & Schuster, 1986), 15.

39. J. D. Bernal, *The World, the Flesh, and the Devil: An Enquiry into the Future of the Three Enemies of the Rational Soul* (New York: E. P. Dutton, 1929; 2d ed., Bloomington: Indiana University Press, 1969), 28 (page citation is to the 2d edition).

40. Loren Eiseley, "Is Man Alone in Space?" *Scientific American* 189, no. 7 (July 1953): 84.

41. Hobbes, *Leviathan,* 396.

CHAPTER 2

1. Samuel Butler, "Darwin Among the Machines," Canterbury *Press,* 13 June 1863; reprinted in Henry Festing Jones, ed., *Canterbury Settlement and other Early Essays,* vol. 1 of *The Shrewsbury Edition of the Works of Samuel Butler* (London: Jonathan Cape, 1923), 208–210.

2. Samuel Butler, *A First Year in Canterbury Settlement* (London: Longman & Green, 1863); reprinted in Jones, *Canterbury Settlement,* 82.

3. Ibid., 97.

4. Ibid., 106.

5. Samuel Butler, note, June 1887, in Henry Festing Jones, ed., *Samuel Butler: A Memoir (1835–1902),* vol. 1 (London: Macmillan, 1919), 155.

6. Samuel Butler, note, 1901, in Jones, *Samuel Butler,* vol. 1, 158.

7. Jones, *Samuel Butler,* vol. 1, 155.

8. Samuel Butler, "Analysis of Sales, 28 November 1899," in Jones, *Samuel Butler*, vol. 2, 311.

9. Jones, *Samuel Butler*, vol. 1, 273.

10. Sir Joshua Strange Williams to Henry Festing Jones, 19 August 1912, in Jones, *Samuel Butler*, vol. 1, 84.

11. Robert B. Booth, *Five Years in New Zealand* (London: privately printed, 1912), chap. 14; in Jones, *Samuel Butler*, vol. 1, 87.

12. Samuel Butler to O. T. J. Alpers, 17 February 1902, in Jones, *Samuel Butler*, vol. 2, 382.

13. Thomas Huxley to Charles Darwin, 3 February 1880, in Nora Barlow, ed., *The Autobiography of Charles Darwin, 1809–1882: with Original Omissions Restored, edited with Appendix and Notes by his Grand-daughter* (New York: Harcourt Brace, 1958), 211.

14. Jones, *Samuel Butler*, vol. 1, 300.

15. Samuel Butler, *Luck, or Cunning, as the main means of Organic Modification? An attempt to throw additional light upon Darwin's theory of Natural Selection* (London: Trübner & Co., 1887); reprinted as vol. 8 of *The Shrewsbury Edition of the Works of Samuel Butler* (London: Jonathan Cape, 1924), 61.

16. Erasmus Darwin, *Zoonomia; or, The Laws of Organic Life,* vol. 1 (London: J. Johnson, 1794) 505.

17. Ibid., 2.

18. Ibid., 507.

19. Erasmus Darwin, *Zoonomia,* 3d ed., vol. 2 (London; J. Johnson, 1801), 295, 304.

20. Darwin, *Zoonomia,* vol. 1 (1794), 519.

21. Ibid., 524, 527.

22. Ibid., 503.

23. Erasmus Darwin, *The Temple of Nature; or, the Origin of Society: A Poem with Philosophical Notes* (London: J. Johnson, 1803), 119.

24. Darwin, *Zoonomia,* vol. 1, 509.

25. *Monthly Magazine* 13 (1802): 458; quoted in Desmond King-Hele, *Erasmus Darwin* (New York: Scribner's, 1963), 14.

26. Francis Darwin, *The Life and Letters of Charles Darwin, Including an Autobiographical Chapter,* vol. 1 (New York: Appleton & Co., 1896), 6.

27. Erasmus Darwin to Matthew Boulton, 1781, in Desmond King-Hele, "The Lunar Society of Birmingham," *Nature* 212 (15 October 1966): 232.

28. Erasmus Darwin to Matthew Boulton, ca. 1764, in Robert E. Schofield, *The Lunar Society of Birmingham: A Social History of Provincial Science and*

Industry in Eighteenth-Century England (Oxford: Oxford University Press, 1963), 29–30, and Desmond King-Hele, ed., *The Letters of Erasmus Darwin* (Cambridge: Cambridge University Press, 1981), 27–31.

29. Percy Shelley, preface to Mary Wollstonecraft Shelley's *Frankenstein; or, the Modern Prometheus* (London: Lockington, Hughes, Harding, Mayor & Jones, 1818), vii.

30. Mary W. Shelley, introduction to the Standard Novels edition of *Frankenstein; or, the Modern Prometheus* (London: Colburn & Bentley, 1831; reprint, Penguin Classics, 1985), 8 (page citation is to the reprint edition).

31. Erasmus Darwin to Georgiana, duchess of Devonshire, November 1800, in King-Hele, *Letters of Erasmus Darwin*, 325.

32. *Aris's Birmingham Gazette*, 23 October 1762, excerpted in John A. Langford, *A Century of Birmingham Life*, vol. 1 (Birmingham: E. C. Osborne, 1868), 148; as quoted in Schofield, *Lunar Society*, 26.

33. Samuel Coleridge, 27 January 1796, in Earl Leslie Griggs, ed., *Collected letters*, vol. 1 (Oxford: Clarendon Press, 1956), 99.

34. King-Hele, *Erasmus Darwin*, 3.

35. Charles Darwin to Thomas Huxley, in Francis Darwin, ed., *More Letters of Charles Darwin*, vol. 1 (London: John Murray, 1903), 125.

36. Samuel Butler, *Evolution, Old and New; or, The theories of Buffon, Dr. Erasmus Darwin and Lamarck, as compared with that of Charles Darwin* (London: Hardwicke & Bogue, 1879).

37. Ernst Krause, *Life of Erasmus Darwin, with a Preliminary Notice by Charles Darwin* (London: Charles Murray, 1879), excerpted in Samuel Butler, *Unconscious Memory* (London: David Bogue, 1880; reprint, London: Jonathan Cape, 1924), 42 (page citation is to the reprint edition).

38. Samuel Butler, "Barrel-Organs," Canterbury *Press*, 17 January 1863; reprinted in Jones, *Canterbury Settlement*, 196. Butler ascribed this anonymous letter to Bishop Abraham of Wellington; there is reason to believe he planted it himself.

39. Thomas Butler, in Francis Darwin, *Letters of Charles Darwin*, vol. 1, 144.

40. Charles Darwin, 1876, "Autobiography," in Francis Darwin, *Letters of Charles Darwin*, vol. 1, 29.

41. Samuel Butler, *Unconscious Memory* (London: David Bogue, 1880); reprinted as vol. 6 of *The Shrewsbury Edition of the Works of Samuel Butler* (London: Jonathan Cape, 1924), 4.

42. Ibid., 12.

43. Charles Darwin, 24 March 1863; quoted in Henry Festing Jones, "Darwin on the Origin of Species: Prefatory Note," in Jones, *Canterbury Settlement*, 184–185.

44. Butler, "Darwin Among the Machines," 208.

45. Samuel Butler, "The Mechanical Creation," *Reasoner* (London), 1 July 1865; reprinted in Jones, *Canterbury Settlement*, 231–233.

46. Butler, *Luck, or Cunning?*, 120.

47. Thomas Huxley, 1870, "On Descartes 'Discourse touching the method of using one's reason rightly and of seeking scientific truth,'" reprinted in *Methods and Results*, vol. 1 of *Essays* (New York: Appleton, 1902), 191.

48. Samuel Butler, *Erewhon; or, Over the Range* (London: Trübner & Co., 1872; new and rev. ed., London: A. C. Fifield, 1913), 236–241 (page citations are to the revised edition).

49. Butler to Darwin, 11 May 1872, in Jones, *Samuel Butler*, vol. 1, 156–157.

50. Butler to Darwin, 30 May 1872, in Jones, *Samuel Butler*, vol. 1, 158.

51. Charles Darwin to Thomas Huxley, 4 February 1880, in Jones, *Samuel Butler*, vol. 2, 454.

52. Henry Festing Jones, *Charles Darwin and Samuel Butler: A step towards Reconciliation* (London: A.C. Fifield, 1911); reprinted as an appendix to Barlow, *Autobiography of Charles Darwin*, 174–196.

53. Review of Samuel Butler's *Evolution, Old and New, Saturday Review* (London) 47, no. 1,231 (31 May 1879): 682.

54. Butler, *Unconscious Memory*, 53, 56.

55. Samuel Butler to Thomas Gale Butler, 18 February 1876, in H. F. Jones, ed., *The Notebooks of Samuel Butler* (London: A.C. Fifield, 1912); reprinted as vol. 20 of *The Shrewsbury Edition of the Works of Samuel Butler* (London: Jonathan Cape, 1926), 48.

56. Butler, *Luck, or Cunning?*, 1.

57. Butler, *Unconscious Memory*, 13, 15.

58. Ibid., 13.

59. Freeman J. Dyson, *Origins of Life* (Cambridge: Cambridge University Press, 1985), 8–9.

60. Freeman J. Dyson, "A Model for the Origin of Life," *Journal of Molecular Evolution* 18 (1982): 344.

61. Freeman J. Dyson, *Collected Scientific Papers with Commentary* (Providence, R.I.: American Mathematical Society, 1996), 47.

62. Dyson, *Origins of Life*, 5.

63. Butler, "Mechanical Creation," 233.

64. Butler, *Erewhon*, 252–255.

65. Thomas Huxley, 1887, "The Progress of Science," reprinted in *Methods and Results*, 117.

66. Dyson, *Origins of Life,* 7.

67. Samuel Butler, "From our Mad Correspondent," Canterbury *Press,* 15 September 1863; reprinted in Joseph Jones, *The Cradle of Erewhon: Samuel Butler in New Zealand* (Austin: University of Texas Press, 1959), 196–197.

68. Samuel Butler, "Lucubratio Ebria," Canterbury *Press,* 29 July 1865; reprinted in Jones, *Notebooks of Samuel Butler,* 40.

69. Butler, *Unconscious Memory,* 57.

CHAPTER 3

1. Charles Babbage, *The Ninth Bridgewater Treatise: A Fragment,* 2d ed. (London: John Murray, 1838), 33.

2. Leibniz to Hobbes, 13/23 July 1670, in Noel Malcolm, ed., *The Correspondence of Thomas Hobbes,* vol. 2 (Oxford: Oxford University Press, 1994), 720.

3. Olaf Stapledon, "Interplanetary Man," *Journal of the British Interplanetary Society,* 7, no. 6 (7 November 1948): 231.

4. E. T. Bell, *Men of Mathematics* (New York: Simon & Schuster, 1937), 120, 122.

5. Leibniz to Henry Oldenburg, 18 December 1675, in H. W. Turnbull, ed., *The Correspondence of Isaac Newton,* vol. 1 (Cambridge: Cambridge University Press, 1959), 401.

6. Leibniz to Nicolas Remond, 10 January 1714, in Leroy E. Loemker, trans. and ed., *Philosophical Papers and Letters,* vol. 2 (Chicago: University of Chicago Press, 1956), 1063.

7. Leibniz, 1685, "Machina arithmetica in qua non additio tantum et subtractio sed et multiplicatio nullo, divisio vero pæne nullo animi labore peragantur," translated as "Leibniz on his Calculating Machine," in D. E. Smith, ed., *A Source Book in Mathematics,* vol. 1 (New York: Dover, 1929), 180.

8. Leibniz, letter, n.d., quoted in H. W. Buxton, 1871, *Memoir of the Life and Labours of the Late Charles Babbage Esq. F.R.S. (MS, 1871),* Charles Babbage Institute Reprint Series for the History of Computing, vol. 13 (Cambridge: MIT Press, 1988), 51, 381.

9. Leibniz, 1685, in Smith, *Source Book,* vol. 1, 180–181.

10. Leibniz, 1716, in Henry Rosemont, Jr., and Daniel J. Cook, trans. and eds., *Discourse on the Natural Theology of the Chinese* (translation of "Lettre sur la philosophie chinoise à Nicolas de Remond"), Monograph of the Society for Asian and Comparative Philosophy, no. 4 (Honolulu: University of Hawaii Press, 1977), 158.

11. Leibniz, "De Progressione Dyadica—Pars I," (MS, 15 March 1679), published in facsimile (with German translation) in Erich Hochstetter and Hermann-Josef Greve, eds., *Herrn von Leibniz' Rechnung mit Null und Eins* (Berlin: Siemens Aktiengesellschaft, 1966), 46–47. (English translation by Verena Huber-Dyson, 1995.)

12. Leibniz, ca. 1679, in Loemker, *Philosophical Papers,* vol. 1, 342.

13. Ibid., 344.

14. Leibniz, supplement to a letter to Christiaan Huygens, 8 September 1679, in Loemker, *Philosophical Papers,* vol. 1, 384–385.

15. Charles Babbage, *Passages from the Life of a Philosopher* (London: Longman, Green, 1864), 142. Facsimile reprint, New York: A. M. Kelley, 1969.

16. Buxton, *Babbage,* 158.

17. Ibid., 155.

18. Leibniz, 1710, "Reflexions on the Work that Mr. Hobbes Published in English on 'Freedom, Necessity and Chance,'" in E. M. Huggard, trans., and Austin Farrer, ed., *Theodicy: Essays on the Goodness of God the Freedom of Man and the Origin of Evil,* (La Salle, Ill.: Open Court, 1951), 393.

19. Babbage, *Passages,* 42.

20. Buxton, *Babbage,* 46.

21. Babbage, *Passages,* 118–119.

22. Doron D. Swade, "Redeeming Charles Babbage's Mechanical Computer," *Scientific American* 268, no. 2 (February 1993): 86.

23. Charles Darwin, 1876, in Nora Barlow, ed., *The Autobiography of Charles Darwin, 1809–1882: with Original Omissions Restored, edited with Appendix and Notes by his Grand-daughter* (New York: Harcourt Brace, 1958), 108. This reference to Babbage, and accompanying comments on Herbert Spencer, were deleted from the version published by Francis Darwin in 1896.

24. Ada Augusta Lovelace, Note A to L. F. Menabrea's "Sketch of the Analytical Engine invented by Charles Babbage, Esq.," *Taylor's Scientific Memoirs,* vol. 3 (London: J. E. & R. Taylor, 1843), reprinted in Henry Provost Babbage, ed., *Babbage's Calculating Engines: Being a Collection of Papers Relating to them; their History, and Construction* (London: E. and F. Spon, 1889), 25. Facsimile reprint, Charles Babbage Institute Reprint Series for the History of Computing, vol. 2 (Cambridge, Mass.: MIT Press, 1982).

25. Babbage, *Ninth Bridgewater Treatise,* 97.

26. Ibid., vii.

27. Charles Babbage, *On the Economy of Machinery and Manufactures,* 4th ed., enlarged (London: Charles Knight, 1835), 273–276.

28. Babbage, *Passages,* 128.

29. George Boole, *An Investigation of the Laws of Thought, on which are founded the mathematical theories of Logic and Probabilities* (London: Macmillan, 1854), 1.

30. Herman Goldstine, *The Computer from Pascal to von Neumann* (Princeton, N.J.: Princeton University Press, 1972), 153.

31. John von Neumann, "Probabilistic Logics and the Synthesis of Reliable Organisms from Unreliable Components," in Claude E. Shannon and John McCarthy, eds., *Automata Studies* (Princeton, N.J.: Princeton University Press, 1956), 43–99.

32. Boole, *Laws of Thought*, 21.

33. Ibid., 408.

34. Leibniz, ca. 1702, "Reflections on the Common Concept of Justice," in Loemker, *Philosophical Papers*, vol. 2, 919.

35. D'arcy Power, in *Dictionary of National Biography*, vol. 18 (London: Smith, Elder & Co., 1898), 399.

36. Alfred Smee, *Principles of the Human Mind deduced from Physical Laws* (London: Longman, Brown, Green, & Longmans, 1849); reprinted in Elizabeth Mary (Smee) Odling, *Memoir of the late Alfred Smee, F. R. S., by his daughter; with a selection from his miscellaneous writings* (London: George Bell & Sons, 1878), 271.

37. Alfred Smee, *The Process of Thought Adapted to Words and Language, together with a description of the Relational and Differential Machines* (London: Longman, Brown, Green, & Longmans, 1851), ix.

38. Ibid., 2.

39. Ibid., 25.

40. Ibid., 39.

41. Ibid., 42–43.

42. Ibid., 48–49.

43. Ibid., 49–50.

44. Alfred Smee, *Instinct and Reason: Deduced from Electro-Biology* (London: Reeve, Benham & Reeve, 1850), 97.

45. Alfred Smee, *Elements of Electro-Biology; or, the Voltaic Mechanism of Man; of Electro-Pathology, Especially of the Nervous System; and of Electro-Therapeutics* (London: Reeve, Benham & Reeve, 1849), 20.

46. Smee, *Instinct and Reason*, 28–29.

47. Ibid., 200, 221.

48. *Saturday Review* (London), 10 August 1872, 194.

49. Power, *Dictionary*, vol. 18, 399.

50. Kurt Gödel, "Über formal unentscheidbare Sätze der *Principia Mathematica* und verwandter Systeme I," *Monatshefte für Mathematik und Physik* 38 (1931); translated by Elliott Mendelson as "On Formally Undecidable Propositions of *Principia Mathematica* and Related Systems I," in Martin Davis, ed., *The Undecidable* (Hewlett, N.Y.: Raven Press, 1965), 5.

51. Leibniz to Clarke, 18 August 1716, in H. G. Alexander, ed., *The Leibniz–Clarke Correspondence* (Manchester, England: Manchester University Press, 1956), 193.

52. Leibniz, 1714, *The Monadology,* in George R. Montgomery, trans., *Basic Writings: Discourse on Metaphysics; Correspondence with Arnauld; Monadology* (La Salle, Ill.: Open Court, 1902), 254.

53. Leibniz to Caroline, Princess of Wales, ca. 1716, in Alexander, *Correspondence,* 191.

CHAPTER 4

1. Alan Turing, "Computing Machinery and Intelligence," *Mind* 59 (October 1950): 443.

2. A. K. Dewdney, *The Turing Omnibus* (Rockville, Md.: Computer Science Press, 1989), 389.

3. Robin Gandy, "The Confluence of Ideas in 1936," in Rolf Herken, ed., *The Universal Turing Machine: A Half-century Survey* (Oxford: Oxford University Press, 1988), 85.

4. Alan Turing, "On Computable Numbers, with an Application to the Entscheidungsproblem," *Proceedings of the London Mathematical Society,* 2d ser. 42 (1936–1937); reprinted, with corrections, in Martin Davis, ed., *The Undecidable* (Hewlett, N.Y.: Raven Press, 1965), 117.

5. Ibid., 136.

6. Kurt Gödel, 1946, "Remarks Before the Princeton Bicentennial Conference on Problems in Mathematics," reprinted in Davis, *The Undecidable,* 84.

7. W. Daniel Hillis, *The Difference That Makes a Difference* (New York: Basic Books, forthcoming).

8. Malcolm MacPhail to Andrew Hodges, 17 December 1977, in Andrew Hodges, *Alan Turing: The Enigma* (New York: Simon & Schuster, 1983), 138.

9. Allan Marquand, "A New Logical Machine," *Proceedings of the American Academy of Arts and Sciences* 21 (1885): 303.

10. Charles Peirce to Allan Marquand, 1866, in Arthur W. Burks, "Logic, Computers, and Men," *Proceedings and Addresses of the American Philosophical Association* 46 (1973): 47–48.

11. Wolfe Mays, "The First Circuit of an Electrical Logic-Machine," *Science* 118 (4 September 1953): 281.

12. George W. Patterson, "The First Electric Computer, a Magnetological Analysis," *Journal of the Franklin Institute* 270 (1960): 130.

13. Charles S. Peirce, "Logical Machines," *American Journal of Psychology* 1 (November 1887): 165.

14. Ibid., 170.

15. Ibid., 168.

16. Ibid., 169.

17. Theodosia Talcott to H. Talcott, 6 January 1889, in Geoffrey D. Austrian, *Herman Hollerith: Forgotten Giant of Information Processing* (New York: Columbia University Press, 1982), 39–40.

18. Emmanuel Scheyer, "When Perforated Paper Goes to Work: How Strips of Paper Can Endow Inanimate Machines with Brains of Their Own," *Scientific American* 127 (December 1922): 395.

19. Vannevar Bush, "Instrumental Analysis," *Bulletin of the American Mathematical Society* 42 (October 1936): 652.

20. John W. Tukey, 9 January 1947, "Sequential Conversion of Continuous Data to Digital Data," in Henry S. Tropp, "Origin of the Term Bit," *Annals of the History of Computing* 6, no. 2 (April 1984): 153–154.

21. Claude E. Shannon, "A Mathematical Theory of Communication," *Bell System Technical Journal* 27 (July and October 1948): 379–423, 623–656.

22. Bush, "Instrumental Analysis," 653–654.

23. Irving J. Good, "Pioneering Work on Computers at Bletchley," in Nicholas Metropolis, J. Howlett, and Gian-Carlo Rota, eds., *A History of Computing in the Twentieth Century* (New York: Academic Press, 1980), 35.

24. Peter Hilton, "Reminiscences of Bletchley Park, 1942–1945," in *A Century of Mathematics in America,* part 1 (Providence, R.I.: American Mathematical Society, 1988), 293–294.

25. Diana Payne, "The Bombes," in F. H. Hinsley and Alan Stripp, eds., *Codebreakers: The Inside Story of Bletchley Park* (Oxford: Oxford University Press, 1993), 134.

26. Thomas H. Flowers, "The Design of Colossus," *Annals of the History of Computing* 5 (1983): 244.

27. Irving J. Good, "A Report on a Lecture by Tom Flowers on the Design of the Colossus," *Annals of the History of Computing* 4, no. 1 (1982): 57–58.

28. Howard Campaigne, introduction to Flowers, "Design of Colossus," 239.

29. Irving J. Good, "Enigma and Fish," revised, with corrections, in F. H. Hinsley and Alan Stripp, eds., *Codebreakers: The Inside Story of Bletchley Park*, 2d ed. (Oxford: Clarendon Press, 1994), 164.

30. Hodges, *Turing,* 278.

31. Irving J. Good, "Turing and the Computer," review of *Alan Turing: The Enigma,* by Andrew Hodges, *Nature* 307 (1 February 1984): 663.

32. Brian Randell, "The Colossus," in Metropolis, Howlett, and Rota, *History of Computing,* 78.

33. Hilton, "Reminiscences," 293.

34. Alan Turing, "Proposal for the Development in the Mathematics Division of an Automatic Computing Engine (ACE)," reprinted in B. E. Carpenter and R. W. Doran, eds., *A. M. Turing's A.C.E. Report of 1946 and Other Papers,* Charles Babbage Reprint Series for the History of Computing, vol. 10 (Cambridge: MIT Press, 1986), 20–105.

35. Hodges, *Turing,* 307.

36. Carpenter and Doran, *Turing's A.C.E. Report,* 2.

37. Sara Turing, *Alan M. Turing* (Cambridge: W. Heffer & Sons, 1959), 78.

38. M. H. A. Newman, quoted in Good, "Turing and the Computer," 663.

39. Alan Turing, "Lecture to the London Mathematical Society on 20 February 1947," in Carpenter and Doran, *Turing's A.C.E. Report,* 112.

40. Ibid., 106.

41. J. H. Wilkinson, "Turing's Work at the National Physical Laboratory," in Metropolis, Howlett, and Rota, *History of Computing,* 111.

42. Alan Turing, "Intelligent Machinery," report submitted to the National Physical Laboratory, 1948, in Donald Michie, ed., *Machine Intelligence,* vol. 5 (1970), 3.

43. Turing, "Lecture," 124.

44. Turing, "Intelligent Machinery," 4.

45. Turing, "Lecture," 123.

46. Turing, "Intelligent Machinery," 9.

47. Ibid., 23.

48. Turing, "Computing Machinery," 456.

49. Turing, "Intelligent Machinery," 21–22.

50. Turing, "Systems of Logic Based on Ordinals," *Proceedings of the London Mathematical Society,* 2d ser. 45 (1939); reprinted in Davis, *The Undecidable,* 209.

51. John von Neumann, 1948, "The General and Logical Theory of Automata," in Lloyd A. Jeffress, ed., *Cerebral Mechanisms in Behavior: The Hixon Symposium* (New York: Hafner, 1951), 26.

52. Leibniz, 1714, *The Monadology,* in George R. Montgomery, trans., *Basic Writings: Discourse on Metaphysics; Correspondence with Arnauld; Monadology* (La Salle, Ill.: Open Court, 1902), 253.

CHAPTER 5

1. John von Neumann to Gleb Wataghin, ca. 1946, as reported by Freeman J. Dyson, *Disturbing the Universe* (New York: Harper & Row, 1979), 194.

2. Stanislaw Ulam, *Adventures of a Mathematician* (New York: Scribner's, 1976), 231.

3. Nicholas Vonneumann, "John von Neumann: Formative Years," *Annals of the History of Computing* 11, no. 3 (1989): 172.

4. Eugene P. Wigner, "John von Neumann—A Case Study of Scientific Creativity," *Annals of the History of Computing* 11, no. 3 (1989): 168.

5. Edward Teller, in Jean R. Brink and Roland Haden, "Interviews with Edward Teller and Eugene P. Wigner," *Annals of the History of Computing* 11, no. 3 (1989): 177.

6. Stanislaw Ulam, "John von Neumann, 1903–1957," *Bulletin of the American Mathematical Society* 64, no. 3 (May 1958): 1.

7. Eugene Wigner, "Two Kinds of Reality," *The Monist* 49, no. 2 (April 1964); reprinted in *Symmetries and Reflections* (Cambridge: MIT Press, 1967), 198.

8. John von Neumann, statement on nomination to membership in the AEC, 8 March 1955, von Neumann Papers, Library of Congress; in William Aspray, *John von Neumann and the Origins of Modern Computing* (Cambridge: MIT Press, 1990), 247.

9. John von Neumann, as quoted by J. Robert Oppenheimer in testimony before the AEC Personnel Security Board, 16 April 1954, *In the Matter of J. Robert Oppenheimer* (Washington, D.C.: Government Printing Office, 1954; reprint, Cambridge: MIT Press, 1970), 246 (page citation is to the reprint edition).

10. Nicholas Metropolis, "The MANIAC," in Nicholas Metropolis, J. Howlett, and Gian-Carlo Rota, eds., *A History of Computing in the Twentieth Century* (New York: Academic Press, 1980), 459.

11. John von Neumann, testimony before the AEC Personnel Security Board, 27 April 1954, *In the Matter of J. Robert Oppenheimer*, 655.

12. Ralph Slutz, interview by Christopher Evans, June 1976, OH 086, Charles Babbage Institute, University of Minnesota, Minneapolis.

13. John von Neumann, "The Role of Mathematics in the Sciences and Society," address to Princeton graduate alumni, June 1954; reprinted in John von Neumann, *Theory of Games, Astrophysics, Hydrodynamics and Meteorology*, vol. 6 of *Collected Works* (Oxford: Pergamon Press, 1963), 478, 490.

14. Galileo Galilei, *Dialogues Concerning Two New Sciences* (Leyden: Elzevir, 1638), 275; trans. Henry Crew and Alfonso De Salvio (New York: Macmillan, 1914; reprint, Evanston, Ill.: Northwestern University Press, 1946), 242 (page citation is to the reprint edition).

15. Ibid., 246.

16. Herman Goldstine, 16 August 1944, in *The Computer from Pascal to von Neumann* (Princeton, N.J.: Princeton University Press, 1972), 166.

17. William H. Calvin, "A Stone's Throw and Its Launch Window: Timing Precision and Its Implications for Language and Hominid Brains," *Journal of Theoretical Biology* 104 (September 1983): 121.

18. Robert Oppenheimer to James Conant, October 1949, AEC Records; in James R. Shepley and Clay Blair, *The Hydrogen Bomb* (Westport, Conn.: Greenwood Press, 1954), 70.

19. Willis H. Ware, *The History and Development of the Electronic Computer Project at the Institute for Advanced Study,* RAND Corporation Memorandum P-377, 10 March 1953, 5–6.

20. Martin Schwarzschild, interview by William Aspray, 18 November 1986, OH 124, Charles Babbage Institute, University of Minnesota, Minneapolis.

21. Richard Feynman, "Los Alamos from Below—Reminiscences of 1943–1945," *Engineering and Science* 39, no. 2 (January–February 1976): 25.

22. Osborne Reynolds, "An Experimental Investigation of the Circumstances which determine whether the Motion of Water shall be direct or sinuous, and the Laws of Resistance in parallel Channels," *Philosophical Transactions of the Royal Society of London* 174 (1883): 936.

23. Ibid., 938.

24. Stanislaw Ulam, "Von Neumann: The Interaction of Mathematics and Computing," in Metropolis, Howlett, and Rota, *History of Computing,* 93.

25. Norbert Wiener, *I Am a Mathematician* (New York: Doubleday, 1956), 260.

26. Lewis Fry Richardson, *Weather Prediction by Numerical Process* (Cambridge: Cambridge University Press, 1922; facsimile reprint, New York: Dover Publications, 1965), xiii.

27. Ibid., xi.

28. W. Daniel Hillis, "Richard Feynman and the Connection Machine," *Physics Today* 42, no. 2 (1989): 78.

29. Lewis Fry Richardson, *Arms and Insecurity: A Mathematical Investigation into the Causes of War,* ed. Quincy Wright and C. C. Lienau (Pittsburgh: The Boxwood Press, 1960); *Statistics of Deadly Quarrels,* ed. Quincy Wright and C. C. Lienau (Pittsburgh: The Boxwood Press, 1960).

30. Lewis Fry Richardson, "The Analogy Between Mental Images and Sparks," *Psychological Review* 37, no. 3 (May 1930): 222.

31. Sidney Shalett, "Electronics to Aid Weather Figuring," *New York Times,* 11 January 1946, 12.

32. Stanislaw Ulam, *Science, Computers and People: From the Tree of Mathematics* (Boston: Birkhauser, 1986), 164.

33. Shalett, "Electronics," 12.

34. Stan Frankel, letter to Brian Randell, 1972, in Brian Randell, "On Alan Turing and the Origins of Digital Computers," *Machine Intelligence* 7 (1972): 10.

35. Rudolf Ortvay to John von Neumann, Budapest, 29 January 1941, in Denes Nagy, ed., "The von Neumann–Ortvay Connection," *Annals of the History of Computing* 11, no. 3 (1989): 187.

36. Warren S. McCulloch and Walter Pitts, "A Logical Calculus of the Ideas Immanent in Nervous Activity," *Bulletin of Mathematical Biophysics* 5 (1943): 115–133.

37. John von Neumann, n.d., Library of Congress, summarized in Aspray, *von Neumann*, 271.

38. Herman H. Goldstine, interview by Nancy Stern, 11 August 1980, OH 018, Charles Babbage Institute, University of Minnesota, Minneapolis.

39. Metropolis, Howlett, and Rota, *History of Computing*, xvii.

CHAPTER 6

1. Arthur Burks, Herman Goldstine, and John von Neumann, 1946, *Preliminary Discussion of the Logical Design of an Electronic Computing Instrument* (Princeton, N.J.: Institute for Advanced Study, 28 June 1946; 2d ed., September 1947); reprinted in John von Neumann, *Design of Computers, Theory of Automata and Numerical Analysis*, vol. 5 of *Collected Works*, ed. Abraham Taub (Oxford: Pergamon Press, 1963), 79.

2. Harry Woolf, ed., *A Community of Scholars: The Institute for Advanced Study Faculty and Members, 1930–1980* (Princeton, N.J.: Institute for Advanced Study, 1980), ix.

3. Ibid., 130.

4. Abraham Flexner, *I Remember* (New York: Simon & Schuster, 1940), 13.

5. Abraham Flexner, "The Usefulness of Useless Knowledge," *Harper's Magazine*, October 1939, 548.

6. Flexner, *I Remember*, 75.

7. Ibid., 356.

8. Flexner, "Useless Knowledge," 551.

9. Flexner, *I Remember*, 361, 375.

10. Flexner, "Useless Knowledge," 551.

11. Ibid., 552.

12. Flexner, *I Remember*, 375.

13. Ibid., 377–378.

14. Flexner, "Useless Knowledge," 551.

15. Ibid., 551.

16. Flexner, *I Remember*, 375.

17. Arthur W. Burks, interview by William Aspray, 20 June 1987, OH 136, Charles Babbage Institute, University of Minnesota, Minneapolis.

18. Willis H. Ware, interview by Nancy Stern, 19 January 1981, OH 37, Charles Babbage Institute, University of Minnesota, Minneapolis.

19. John von Neumann, "Governed," review of *Cybernetics,* by Norbert Wiener, *Physics Today* 2 (1949): 33.

20. Willis H. Ware, *The History and Development of the Electronic Computer Project at the Institute for Advanced Study,* RAND Corporation Memorandum P-377, 10 March 1953, 7–8.

21. Burks, interview.

22. John von Neumann, "Memorandum on the Program of the High-Speed Computer," 8 November 1945, quoted in Herman Goldstine, *The Computer from Pascal to von Neumann* (Princeton, N.J.: Princeton University Press, 1972), 255.

23. Irving J. Good, "Some Future Social Repercussions of Computers," *International Journal of Environmental Studies* 1 (1970): 69.

24. Burks, interview.

25. Ralph Slutz, interview by Christopher Evans, June 1976, OH 86, Charles Babbage Institute, University of Minnesota, Minneapolis.

26. Ware, interview.

27. Herman H. Goldstine, interview by Nancy Stern, 11 August 1980, OH 18, Charles Babbage Institute, University of Minnesota, Minneapolis.

28. Norbert Wiener, *I Am a Mathematician* (New York: Doubleday, 1956), 242–243.

29. Julian Bigelow, Arturo Rosenblueth, and Norbert Wiener, "Behavior, Purpose and Teleology," *Philosophy of Science* 10, no. 1 (1943): 22.

30. Warren S. McCulloch, "The Imitation of One Form of Life by Another— Biomimesis," in Eugene E. Bernard and Morley R. Kare, eds., *Biological Prototypes and Synthetic Systems,* Proceedings of the Second Annual Bionics Symposium sponsored by Cornell University and the General Electric Company, Advanced Electronics Center, held at Cornell University, August 30–September 1, 1961, vol. 1 (New York: Plenum Press, 1962), 393.

31. Ware, interview.

32. Ibid.

33. Julian Bigelow, "Computer Development at the Institute for Advanced Study," in Nicholas Metropolis, J. Howlett, and Gian-Carlo Rota, eds., *A History of Computing in the Twentieth Century* (New York: Academic Press, 1980), 291.

34. Ware, interview.

35. Burks, interview.

36. Bigelow, "Computer Development," 304.

37. Ibid., 307.

38. Ibid., 297.

39. Ibid., 308.

40. Ibid., 306.

41. William F. Gunning, *Rand's Digital Computer Effort,* Rand Corporation Memorandum P-363, 23 February 1953, 4.

42. Richard W. Hamming, "The History of Computing in the United States, " in Dalton Tarwater, ed., *The Bicentennial Tribute to American Mathematics, 1776–1976* (Washington, D.C.: Mathematical Association of America, 1977), 119.

43. Martin Schwarzschild, interview by William Aspray, 18 November 1986, OH 124, Charles Babbage Institute, University of Minnesota, Minneapolis.

44. Edmund C. Berkeley, *Giant Brains* (New York: John Wiley, 1949), 5.

45. John von Neumann, 1948, "The General and Logical Theory of Automata," in Lloyd A. Jeffress, ed., *Cerebral Mechanisms in Behavior: The Hixon Symposium* (New York: Hafner, 1951), 31.

46. Stanislaw Ulam, *Adventures of a Mathematician* (New York: Scribner's, 1976), 242.

47. John von Neumann, 1948, response to W. S. McCulloch's paper "Why the Mind Is in the Head," Hixon Symposium, September 1948, in Jeffress, *Cerebral Mechanisms,* 109–111.

48. John von Neumann to Oswald Veblen, memorandum, 26 March 1945, "On the Use of Variational Methods in Hydrodynamics," reprinted in John von Neumann, *Theory of Games, Astrophysics, Hydrodynamics and Meteorology,* vol. 6 of *Collected Works,* ed. Abraham Taub (Oxford: Pergamon Press, 1963), 357.

CHAPTER 7

1. Marvin Minsky, 1971, in Carl Sagan, ed., *Communication with Extraterrestrial Intelligence,* Proceedings of the Conference held at the Byurakan Astrophysical Observatory, Yerevan, USSR, 5–11 September 1971 (Cambridge: MIT Press, 1973), 328.

2. Julian Bigelow, "Computer Development at the Institute for Advanced Study," in Nicholas Metropolis, J. Howlett, and Gian-Carlo Rota, eds., *A History of Computing in the Twentieth Century* (New York: Academic Press, 1980), 308.

3. Konstantin S. Merezhkovsky, *Theory of Two Plasms as the Basis of Symbiogenesis: A New Study on the Origin of Organisms* (in Russian) (Kazan: Publishing Office of the Imperial Kazan University, 1909); Boris M. Kozo-Polyansky, *A New Principle of Biology: Essay on the Theory of Symbiogenesis* (in Russian) (Moscow, 1924). The theory is most accessible in English in Liya N. Khakhina's *Concepts of Symbiogenesis: A Historical and Critical Study of the Research of Russian Botanists,* trans. Stephanie Merkel, ed. Lynn Margulis and Mark McMenamin (New Haven, Conn.: Yale University Press, 1992).

4. Merezhkovsky, *Theory of Two Plasms,* 8; after Khakina, *Symbiogenesis,* ii.

5. Edmund B. Wilson, *The Cell in Development and Heredity,* 3d ed. (New York: Macmillan, 1925), 738.

6. Nils A. Barricelli, "Numerical Testing of Evolution Theories: Part 1," *Acta Biotheoretica* 16 (1962): 94.

7. Nils A. Barricelli, "Numerical Testing of Evolution Theories: Part 2," *Acta Biotheoretica* 16 (1962): 122.

8. Barricelli, "Numerical Testing of Evolution Theories: Part 1," 70.

9. James Pomerene, interview by Nancy Stern, 26 September 1980, OH 31, Charles Babbage Institute, University of Minnesota, Minneapolis.

10. Nils A. Barricelli, "Symbiogenetic Evolution Processes Realized by Artificial Methods," *Methodos* 9, nos. 35–36 (1957): 152.

11. Barricelli, "Numerical Testing of Evolution Theories: Part 1," 72.

12. Barricelli, "Symbiogenetic Evolution Processes," 169.

13. Ibid., 164.

14. Barricelli, "Numerical Testing of Evolution Theories: Part 1," 70.

15. Ibid., 76.

16. Nils A. Barricelli, "Numerical Testing of Evolution Theories," *Journal of Statistical Computation and Simulation* 1 (1972): 123–124.

17. Barricelli, "Numerical Testing of Evolution Theories: Part 1," 94.

18. Barricelli, "Symbiogenetic Evolution Processes," 159.

19. Barricelli, "Numerical Testing of Evolution Theories: Part 1," 89.

20. Ibid., 69, 99.

21. Ibid., 94.

22. Ibid., 73.

23. Barricelli, "Numerical Testing of Evolution Theories: Part 2," 100.

24. Ibid., 116.

25. Barricelli, "Numerical Testing of Evolution Theories: Part 1," 122.

26. Barricelli, "Numerical Testing of Evolution Theories: Part 2," 100.

27. Barricelli, "Numerical Testing of Evolution Theories: Part 1," 126.

28. Barricelli, "Numerical Testing of Evolution Theories: Part 2," 117.

29. A. G. Cairns-Smith, *Seven Clues to the Origin of Life* (Cambridge: Cambridge University Press, 1985), 106.

30. Tor Gulliksen, personal communication, 22 November 1995.

31. Ibid.

32. Simen Gaure, personal communication, 23 November 1995.

33. Nils Barricelli, in Paul S. Moorhead and Martin M. Kaplan, eds., *Mathematical Challenges to the Neo-Darwinian Interpretation of Evolution*, A Symposium Held at the Wistar Institute, April 25–26, 1966 (Philadelphia: Wistar Institute, 1967), 64.

34. Gaure, personal communication.

35. Barricelli, "Numerical Testing of Evolution Theories: Part 2," 101.

36. John Backus, "Programming in America in the 1950s—Some Personal Impressions," in Metropolis, Howlett, and Rota, *History of Computing*, 127.

37. Data in this paragraph are from Montgomery Phister, Jr., *Data Processing Technology and Economics*, 2d ed. (Bedford, Mass.: Digital Press, 1979), 19, 26, 27, 215, 277, 531, 611.

38. Nils A. Barricelli, "The Functioning of Intelligence Mechanisms Directing Biologic Evolution," *Theoretic Papers* 3, no. 7 (1985): 126.

39. Maurice Wilkes, *Memories of a Computer Pioneer* (Cambridge: MIT Press, 1985), 145.

40. Barricelli, "Symbiogenetic Evolution Processes," 147.

41. Thomas Ray, "Evolution, Complexity, Entropy, and Artificial Reality," preprint submitted to *Physica D* (20 August 1993): 2.

42. Thomas Ray, "How I Created Life in a Virtual Universe" (unpublished preprint, School of Life and Health Sciences, University of Delaware, 29 March 1992), 5–6.

43. Ibid., 6.

44. Thomas Ray, "An Evolutionary Approach to Synthetic Biology: Zen and the Art of Creating Life," preprint submitted to *Artificial Life* 1, no. 1 (21 October 1993): 5.

45. Thomas Ray, "A Proposal to Create a Network-Wide Biodiversity Reserve for Digital Organisms," preprint, ATR Human Information Processing Research Laboratories, Kyoto, Japan (2 March 1994), 2.

46. Thomas Ray and Kurt Thearling, "Evolving Multi-cellular Artificial Life," preprint submitted to *Proceedings of Artificial Life IV* (July 1994): 6.

47. Ray, "Proposal," 6.

48. Ibid., 5–6.

49. Ray, "Synthetic Biology," 29.

50. Thomas Ray, "Security," unpublished memo, 1 August 1995.

51. Barricelli, "Numerical Testing of Evolution Theories," 126.

52. Nils Barricelli, "Genetic Language, Its Origins and Evolution," *Theoretic Papers* 4, no. 6 (1986): 106–107.

53. Nils A. Barricelli, "On the Origin and Evolution of the Genetic Code: 2. Origin of the Genetic Code as a Primordial Collector Language; The Pairing-Release Hypothesis," *BioSystems* 11 (1979): 19, 21.

54. Martin Davis, "Influences of Mathematical Logic on Computer Science," in Rolf Herken, ed., *The Universal Turing Machine: A Half-century Survey* (Oxford: Oxford University Press, 1988), 315.

55. Alan Turing, "Computing Machinery and Intelligence," *Mind* 59 (October 1950): 456.

CHAPTER 8

1. W. Daniel Hillis, "New Computer Architectures and Their Relationship to Physics, or Why Computer Science Is No Good," *International Journal of Theoretical Physics* 21, nos. 3–4 (April 1982): 257.

2. Aeschylus, *Agamemnon*, lines 280–316, trans. and ed. Eduard Frankel (Oxford: Clarendon Press, 1950), 109–111.

3. Polybius, *The Histories*, book 10, 45.6–12, trans. W. R. Paton (London: William Heinemann, 1925), 213–214.

4. John Wilkins, *Mercury; or, the Secret and Swift messenger: Shewing, How a Man may with Privacy and Speed communicate his Thoughts to a Friend at any distance* (London: John Maynard, 1641), 88.

5. Ibid., 137.

6. Gerald J. Holzmann and Björn Pehrson, *The Early History of Data Networks* (Los Alamitos, Calif.: IEEE Computer Society Press, 1995), 24.

7. Robert Hooke, 21 May 1684, "Discourse Shewing a Way how to communicate one's Mind at great Distances," in W. Derham, ed., *Philosophical Experiments and Observations of the late Eminent Dr. Robert Hooke* (London: W. Derham, 1726), 142–143.

8. Richard Waller, "The Life of Dr. Robert Hooke," introduction to *The Posthumous Works of Robert Hooke, containing his Cutlerian lectures, and other discourses* (London: Richard Waller, 1705), xxvii.

9. John Aubrey, in *Aubrey's Brief Lives: Edited from the Original Manuscripts with a Life of John Aubrey by Oliver Lawson Dick* (Ann Arbor: University of Michigan Press, 1949), 165.

10. Samuel Pepys, 15 February 1664, in *Diary and Correspondence of Samuel Pepys, F.R.S. . . . Deciphered by Rev. J. Smith, A. M. from the original shorthand MS*, vol. 2 (Philadelphia: John D. Morris, 1890), 211.

11. Waller, "Hooke," ix.

12. Ibid., xiii.

13. Robert Hooke, 7 May 1673, in R. T. Gunther, *Early Science in Oxford*, vol. 7 (Oxford: printed for the author, 1930), 412.

14. Waller, "Hooke," vii.

15. Aubrey, *Brief Lives*, 167.

16. Letter from Hooke to Boyle, 3 July 1663, in Gunther, *Early Science*, vol. 6, 139.

17. Hooke, *Posthumous Works*, 140.

18. Ibid., 144.

19. Journal of the Royal Society, 17 February 1664; in Gunther, *Early Science*, vol. 6, 170.

20. Journal of the Royal Society, 29 February 1672; in Gunther, *Early Science*, vol. 7, 394.

21. Journal of the Royal Society, 7 March 1672; in Gunther, *Early Science*, vol. 7, 394.

22. Hooke, "Discourse," 147.

23. Ibid.

24. Ibid., 146–147.

25. Holzmann and Pehrson, *Data Networks*, 38.

26. Hooke, "Discourse," 148.

27. Gerald J. Holzmann and Björn Pehrson, "The First Data Networks," *Scientific American* 270, no. 1 (January 1994): 129.

28. Abbé Jean Antoine Nollet, *Essai sur l'électricité des corps* (Paris: Frères Guerin, 1746), 135; second quotation in Park Benjamin, *A History of Electricity (The Intellectual rise in electricity) from antiquity to the days of Benjamin Franklin* (New York: John Wiley, 1898), 534.

29. [C.M.], "An expeditious method for conveying intelligence," *Scots' Magazine* 15 (17 February 1745): 73; reprinted in John J. Fahie, *A History of Electric Telegraphy to the Year 1837, chiefly compiled from original sources, and hitherto unpublished documents* (London: E. & F. Spon, 1884), 68–71.

30. Fahie, *Electric Telegraphy*, 221.

31. Francis Ronalds, "Descriptions of an Electrical Telegraph" (London: R. Hunter, 1823), 3; quoted in Fahie, *Electric Telegraphy*, 138.

32. Francis Ronalds to Lord Melville, 11 July 1816; in Fahie, *Electric Telegraphy*, 135.

33. John Barrow to Francis Ronalds, 5 August 1816; in Fahie, *Electric Telegraphy*, 136.

34. André-Marie Ampère, *Recueil d'observations électro-dynamiques contentant divers mémoires, notices, extraits de lettres ou d'ouvrages périodiques sur les sciences, relatifs à l'action mutuelle de deux courans électriques . . .* (Paris: Crochard, 1822), 19. (Author's translation.)

35. John von Neumann, lecture given at University of Illinois, December 1949, in Arthur Burks, ed., *Theory of Self-Reproducing Automata* (Urbana: University of Illinois Press, 1966), 75.

36. John von Neumann, "Defense in Atomic War," *Journal of the American Ordnance Association* (1955): 22; reprinted in John von Neumann, *Theory of Games, Astrophysics, Hydrodynamics and Meteorology*, vol. 6 of *Collected Works* (Oxford: Pergamon Press, 1963), 524.

37. von Neumann, "Defense in Atomic War" (1955), 23; (1963), 525.

38. RAND Articles of Incorporation, 1948, in *The RAND Corporation: The First Fifteen Years* (Santa Monica, Calif.: RAND Corporation, 1963).

39. Contract of 2 March 1946 establishing project RAND; in Bruce Smith, *The RAND Corporation* (Cambridge: Harvard University Press, 1966), 30.

40. *A Million Random Digits with 100,000 Normal Deviates* (Santa Monica, Calif.: RAND Corporation, 1955; reprint, New York: Free Press, 1966), xii (page citation is to the reprint edition).

41. Louis Ridenour and Francis Clauser, *Preliminary Design of an Experimental Earth-Circling Spaceship*, U.S. Air Force Project RAND Report SM-11827, 2 May 1946, 2, 16.

42. RAND, *The RAND Corporation*, 23.

43. Paul Baran, interview by Judy O'Neill, 5 March 1990, OH 182, Charles Babbage Institute, University of Minnesota, Minneapolis.

44. J. M. Chester, *Cost of a Hardened, Nationwide Buried Cable Network*, RAND Corporation Memorandum RM-2627-PR, 1 October 1960.

45. Baran, interview.

46. Ibid.

47. Paul Baran, *Summary Overview*, vol. 11 of *On Distributed Communications*, RAND Corporation Memorandum RM-3767-PR, August 1964, 1.

48. Paul Baran, "Packet Switching," in John C. McDonald, ed., *Fundamentals of Digital Switching*, 2d ed. (New York: Plenum Publishing, 1990), 204.

49. Baran, interview.

50. Paul Baran, *Reliable Digital Communications Systems Utilizing Unreliable Network Repeater Nodes,* RAND Corporation Memorandum P-1995, 27 May 1960, 1–2.

51. Baran, *Digital Communications Systems,* 7.

52. Paul Baran, *History, Alternative Approaches, and Comparisons,* vol. 5 of *On Distributed Communications,* RAND Corporation Memorandum RM-3097-PR, August 1964, 8.

53. Warren S. McCulloch, in Claude Shannon, "Presentation of a Maze-Solving Machine," in Heinz von Foerster, Margaret Mead, and H. L. Teuber, eds., *Cybernetics: Circular, Causal and Feedback Mechanisms in Biological and Social Systems,* Transactions of the Eighth Cybernetics Conference, March 15–16, 1951 (New York: Josiah Macy, Jr., Foundation, 1952); reprinted in N. J. A. Sloane and Aaron D. Wyner eds., *Claude Elwood Shannon: Collected Papers* (New York: IEEE Press, 1993), 687.

54. Baran, *On Distributed Communications,* vol. 5, iii.

55. Baran, "Packet Switching," 209.

56. Baran, *On Distributed Communications,* vol. 1, 25.

57. Ibid., 24.

58. Ibid., 29.

59. Paul Baran, *Security, Secrecy, and Tamper-free Considerations,* vol. 9 of *On Distributed Communications,* RAND Corporation Memorandum RM-3765-PR, August 1964, v.

60. Baran, interview.

61. Ibid.

62. Ibid.

63. Ibid.

CHAPTER 9

1. Stanislaw Ulam, in Paul S. Moorhead and Martin M. Kaplan, eds., *Mathematical Challenges to the Neo-Darwinian Interpretation of Evolution,* A Symposium Held at the Wistar Institute, April 25–26, 1966 (Philadelphia: Wistar Institute, 1967), 42.

2. John von Neumann and Oskar Morgenstern, *Theory of Games and Economic Behavior* (Princeton, N.J.: Princeton University Press, 1944); 2d ed., New York: John Wiley, 1947), 2 (page citation is to the 2d edition).

3. Loren Eiseley, *Darwin's Century* (New York: Doubleday, 1958), 39.

4. André-Marie Ampère, *Considérations sur la théorie mathématique du jeu* (Lyons, France: Frères Perisse, 1802), 3. (Author's translation.)

5. Jacob Marschak, "Neumann's and Morgenstern's New Approach to Static Economics," *Journal of Political Economy* 54, no. 2 (April 1946): 114.

6. J. D. Williams, *The Compleat Strategyst* (Santa Monica, Calif.: RAND Corporation, 1954), 216.

7. John Nash, *Parallel Control*, RAND Corporation Research Memorandum RM-1361, 27 August 1954, 14.

8. John von Neumann, "A Model of General Economic Equilibrium," *Review of Economic Studies* 13 (1945): 1.

9. John von Neumann, *The Computer and the Brain* (New Haven, Conn.: Yale University Press, 1958), 79–82.

10. John von Neumann, 1948, "General and Logical Theory of Automata," in Lloyd A. Jeffress, ed., *Cerebral Mechanisms in Behavior: The Hixon Symposium* (New York: Hafner, 1951), 24.

11. Stan Ulam, quoted by Gian-Carlo Rota, "The Barrier of Meaning," *Letters in Mathematical Physics* 10 (1985): 99.

12. von Neumann, "Automata," 24.

13. Stan Ulam, quoted by Rota, "The Barrier of Meaning," 98.

14. D. E. Rumelhart and J. E. McClelland, *Parallel Distributed Processing: Explorations in the Microstructure of Cognition*, vol. 1 (Cambridge: MIT Press, 1986), 132.

15. William H. Calvin, *The Cerebral Symphony* (New York: Bantam, 1990), 118.

16. Thomas Hobbes, *De Cive* (in Latin) (Paris: privately printed, 1642), chap. 12, part 5; translated by Hobbes as *Philosophicall Rudiments concerning Government and Society* (London: Richard Royston, 1651); reprinted, with an introduction by Sterling Lamprecht, ed., as *De Cive; or, The Citizen* (New York: Appleton-Century-Crofts, 1949), 133.

17. Thomas Hobbes, *Leviathan; or, The Matter, Forme, and Power of a Commonwealth Ecclesiasticall and Civill* (London: Andrew Crooke, 1651), 130–131.

18. John Aubrey, in *Aubrey's Brief Lives: Edited from the Original Manuscripts and with a Life of John Aubrey by Oliver Lawson Dick* (Ann Arbor: University of Michigan Press, 1949), 237.

19. Sir Robert Southwell to William Petty, 28 September 1687, in *The Petty–Southwell Correspondence, 1676–1687, Edited from the Bowood Papers by the Marquis of Landsowne* (London: Constable & Co., 1928), 287.

20. Aubrey, 238.

21. Ibid., 239.

22. Sir William Petty, 7 November 1668, "An attempt to demonstrate that an Engine may be fix'd in a good Ship of 5 or 600 Tonn to give her fresh way at Sea in a calm," in Lord Edmond Fitzmaurice, *The Life of Sir William Petty, 1623–1687* (London: John Murray, 1895), 122–124.

23. William Petty to Robert Southwell, 26 February 1680/81, in *The Petty–Southwell Correspondence*, 87.

24. Lord Shelborne (Charles Petty), dedication to William Petty, *Political Arithmetick; or, a Discourse concerning the extent and value of Lands, People, buildings; Husbandry, Manufacture, Commerce, Fishery, Artizans, Seamen, Soldiers; Public Revenues, Interest, Taxes* . . . (London, 1690).

25. Sir William Petty, 1682, *Quantulumcunque Concerning Money* (London: A. & J. Churchill, 1695), 165.

26. Hilary C. Jenkinson, "Exchequer Tallies," *Archaeologia*, 2d ser., 12 (1911): 368.

27. John Giuseppi, *The Bank of England: A History from Its Foundation in 1694* (Chicago: Henry Regnery Co., 1966), 105.

28. Alfred Smee, *Instinct and Reason: Deduced from Electro-biology* (London: Reeve, Benham & Reeve, 1850), xxix–xxxii.

29. Jenkinson, "Exchequer Tallies," 369.

30. Francis Cradocke, *An Expedient For taking away all Impositions, and raising a Revenue without Taxes, By Erecting Bankes for the Encouragement of Trade* (London: Henry Seile, 1660), 1.

31. Henry Robinson, *Certain Proposals In order to the Peoples Freedome and Accommodation in some Particulars* (London: M. Simmons, 1652), 18.

32. R. L. Rivest, A. Shamir, and L. Adleman, "A Method for Obtaining Digital Signatures and Public-Key Cryptosystems," *Communications of the ACM* 21, no. 2 (February 1978): 120.

33. John Wilkins, *Mercury; or, the Secret and Swift messenger, shewing how a man may with privacy and speed Communicate his thoughts to a Friend at any distance* (London: John Maynard, 1641), 179–180.

34. Ibid., 167.

35. Eric Hughes, "A Long-term Perspective on Electronic Commerce," *Release 1.0* (31 March 1995): 8.

36. Gerald Thompson, "John von Neumann's Contributions to Mathematical Programming Economics," in M. Dore, S. Chakravarty, and Richard Goodwin, eds., *John von Neumann and Modern Economics* (Oxford: Oxford University Press, 1989), 232.

37. Oskar Morgenstern, "The Theory of Games," *Scientific American* 180, no. 5 (May 1949): 23.

38. Marvin Minsky, *The Society of Mind* (New York: Simon & Schuster, 1985), 18, 322.

39. Samuel Butler, *Luck, or Cunning, as the main means of Organic Modification?* (London: Trübner & Co., 1887); reprinted as vol. 8 of *The Shrewsbury Edition of the Works of Samuel Butler* (London: Jonathan Cape, 1924), 98.

40. Adam Smith, 1776, *An Inquiry into the Nature and Causes of the Wealth of Nations,* reprint of the 5th ed., vol. 1 (Chicago: University of Chicago Press, 1904), 477–478.

41. Paul Baran, "Is the UHF Frequency Shortage a Self Made Problem?" address to the Marconi Centennial Symposium, Bologna, Italy, 23 June 1995.

42. Carver Mead, *Analog VLSI and Neural Systems* (Reading, Mass.: Addison-Wesley, 1989), 147.

43. Irving J. Good, 1980, *Ethical Machines* (unpublished draft prepared for the Tenth Machine Intelligence Workshop, Case Western Reserve University, April 20–25, 1981), ix.

44. Irving J. Good, "Speculations Concerning the First Ultraintelligent Machine," *Advances in Computers* 6 (1965): 39–40.

45. William Petty to Robert Southwell, letter, 1677, "The Scale of Creatures," *The Petty Papers: Some Unpublished Writings of Sir William Petty, Edited from the Bowood Papers by the Marquis of Landsowne,* vol. 2 (London: Constable & Co., 1927), 21.

46. W. Stanley Jevons, *Money and the Mechanism of Exchange* (New York: Appleton, 1896), 202.

47. Robert Hooke, *The Posthumous Works of Robert Hooke, containing his Cutlerian lectures, and other discourses* (London: Richard Waller, 1705), 140.

CHAPTER 10

1. Joe Van Lone, Cablevision Inc., quoted by Jerry Michalski in *Release 1.0,* 22 November 1993, 6.

2. Richard Feynman, "There's Plenty of Room at the Bottom," *Engineering and Science* 23 (1960): 26.

3. Ibid., 36.

4. J. B. S. Haldane, "On Being the Right Size," *Possible Worlds* (New York: Harper & Brothers, 1928), 28.

5. W. Ross Ashby, "Principles of the Self-Organizing System," in Heinz von Foerster and George W. Zopf, eds., *Principles of Self-Organization,* Transactions of the University of Illinois Symposium on Self-Organization, 8–9 June 1961 (New York: Pergamon Press, 1962), 266.

6. W. Ross Ashby, "Connectance of Large Dynamic (Cybernetic) Systems: Critical Values for Stability," *Nature* 228 (21 November 1970): 784.

7. W. Ross Ashby, "Principles of the Self-Organizing Dynamic System," *Journal of General Psychology* 37 (1947): 125.

8. W. Ross Ashby, "The Physical Origin of Adaptation by Trial and Error," *Journal of General Psychology* 32 (1945): 24.

9. Ashby, "Trial and Error," 13, 24.

10. Ibid., 20.

11. Ashby, "Principles of the Self-Organizing System," 270, 273.

12. Ibid., 270–271.

13. Irving J. Good, *Speculations on Perceptrons and other Automata,* IBM Research Lecture RC-115 (Yorktown Heights, N.Y.: IBM, 1959), 17.

14. Robert L. Chapman, John L. Kennedy, Allen Newell, and William Biel, "The System Research Laboratory's Air Defense Experiments," *Management Science* 5, no. 3 (April 1959): 260.

15. Ibid., 252.

16. Ibid., 267.

17. John von Neumann, "The Impact of Recent Developments in Science on the Economy and on Economics," speech to the National Planning Association, Washington, D.C., 12 December 1955; reprinted in *Collected Works,* vol. 6 (Oxford: Pergamon Press, 1963), 100.

18. Robert Crago, in "A Perspective on SAGE: Discussion," *Annals of the History of Computing* 5, no. 4 (October 1983): 386.

19. Beatrice K. Rome and Sydney C. Rome, *Leviathan: A Simulation of Behavioral Systems, to Operate Dynamically on a Digital Computer,* System Development Corporation report no. SP-50, 6 November 1959, 7.

20. Ibid., 11.

21. Beatrice K. Rome and Sydney C. Rome, *The Leviathan Technological System for the PHILCO 2000 Computer,* System Development Corporation Technical Memorandum TM-713, 11 April 1962, 8.

22. Rome and Rome, *Leviathan: A Simulation of Behavioral Systems,* 15.

23. Ibid., 24.

24. Ibid., 42.

25. Ibid.

26. Ibid., 48.

27. Beatrice K. Rome and Sydney C. Rome, "Leviathan, and Information Handling in Large Organizations," in Allen Kent and Orrin Taulbee, eds., *Electronic Information Handling* (Washington, D.C.: Spartan Books, 1965), 172–173.

28. Beatrice K. Rome and Sydney C. Rome, *Organizational Growth Through Decisionmaking* (New York: American Elsevier, 1971), 1.

29. Oliver G. Selfridge, "Pandemonium: A Paradigm for Learning," *National Physical Laboratory Symposium No. 10 on the Mechanization of Thought Processes,* vol. 1, proceedings of a symposium held at the National Physical Laboratory, 24–27 November 1958 (London: Her Majesty's Stationery Office, 1959), 516.

30. Ibid., 523.

31. Oliver Selfridge, "Artificial Intelligence and the Future of Software Technology," abstract of lecture sponsored by Barr Systems, Inc., and the Department of Computer and Information Science and Engineering, University of Florida, Gainesville, 23 October 1995.

32. Charles Darwin to Asa Gray, 5 September 1857, in *The Journal and Proceedings of the Linnean Society* 3, no. 9 (1858): 51.

33. Samuel Butler, *Luck, or Cunning, as the main means of Organic Modification?* (London: Trübner & Co., 1887); reprinted as vol. 8 of *The Shrewsbury Edition of the Works of Samuel Butler* (London: Jonathan Cape, 1924), 234.

34. Ibid., 235.

35. Nils Barricelli, "The Intelligence Mechanisms behind Biological Evolution," *Scientia* (September 1963): 178–179.

36. Butler, *Luck, or Cunning?*, 60.

37. Samuel Butler, *Unconscious Memory* (London: David Bogue, 1880); reprinted as vol. 6 of *The Shrewsbury Edition of the Works of Samuel Butler* (London: Jonathan Cape, 1924), 16.

38. William Paley, 1802, *Natural Theology*, vol. 2, reprinted, with illustrative notes, etc., in four volumes (London: Charles Knight, 1845), 9.

39. John von Neumann, in Arthur Burks, ed., *Theory of Self-Reproducing Automata* (Urbana: University of Illinois Press, 1966), 47.

40. John Myhill, "The Abstract Theory of Self-Reproduction," in Mihajlo D. Mesarovic, ed., *Views on General Systems Theory*, Proceedings of the Second Systems Symposium at Case Institute of Technology, 1964; reprinted in Arthur Burks, ed., *Essays on Cellular Automata* (Urbana: University of Illinois Press, 1970), 218.

41. John von Neumann, 1948, "The General and Logical Theory of Automata," in Lloyd A. Jeffress, ed., *Cerebral Mechanisms in Behavior: The Hixon Symposium* (New York: Hafner, 1951), 31.

42. Robert Chambers, *Vestiges of the Natural History of Creation* (London: John Churchill, 1844), 222–223.

43. Nils Barricelli, in Paul S. Moorhead and Martin M. Kaplan, eds., *Mathematical Challenges to the Neo-Darwinian Interpretation of Evolution*, A Symposium Held at the Wistar Institute, April 25–26, 1966 (Philadelphia: Wistar Institute, 1967), 67.

44. Lewis Thomas, "The Lives of a Cell," *New England Journal of Medicine* 284, no. 19 (13 May 1971): 1083.

45. Lewis Thomas, "On Societies as Organisms," *New England Journal of Medicine* 285, no. 29 (8 July 1971): 101.

46. Ibid., 102.

47. Lewis Thomas, "Computers," *New England Journal of Medicine* 288, no. 24 (14 June 1973): 1289.

48. Lewis Thomas, "On Artificial Intelligence," *New England Journal of Medicine*, 28 February 1980: 506.

49. Lewis Thomas, *Late Night Thoughts on Listening to Mahler's Ninth Symphony* (New York: Viking, 1983), 17.

50. Charles Darwin, *The Variation of Animals and Plants under Domestication* (London: John Murray, 1868; 2d ed., New York: Appleton, 1896), 2:399 (page citation is to 2d edition).

51. Philip Morrison, "Entropy, Life, and Communication," in Cyril Ponnamperuma and A. G. W. Cameron, eds., *Interstellar Communication: Scientific Perspectives* (Boston: Houghton Mifflin, 1974), 182.

CHAPTER 11

1. Olaf Stapledon, *Last and First Men* (London: Methuen, 1930); reprinted, from the U.S. edition of 1931, in *Last and First Men & Star Maker* (New York: Dover Publications, 1968), 117.

2. Olaf Stapledon to Agnes Miller, 22 December 1917, in Robert Crossley, ed., *Talking Across the World: The Love Letters of Olaf Stapledon and Agnes Miller, 1913–1919* (Hanover and London: University Press of New England, 1987), 262–263.

3. Olaf Stapledon, "Experiences in the Friends' Ambulance Unit," in Julian Bell, ed., *We Did Not Fight 1914–1918: Experiences of War Resisters* (London: Cobden-Sanderson, 1935), 369.

4. Ibid., 360.

5. Olaf Stapledon to Agnes Miller, 28 February 1915, in Crossley, *Talking Across the World*, 75.

6. Meaburn Tatham and James E. Miles, eds., *The Friends' Ambulance Unit 1914–1919: A Record* (London: Swarthmore Press, 1920), 212.

7. Stapledon, "Experiences," 362.

8. Olaf Stapledon to Agnes Miller, 22 October 1918, in Crossley, *Talking Across the World*, 332.

9. Stapledon, "Experiences," 372.

10. Olaf Stapledon to Agnes Miller, 26 December 1917, in Crossley, *Talking Across the World*, 264–265.

11. Lewis Richardson, as quoted by Ernest Gold, "Lewis Fry Richardson, 1881–1953," *Obituary Notices of Fellows of the Royal Society* 9 (November 1954): 230.

12. Olaf Stapledon to Agnes Miller, 12 January 1918, in Crossley, *Talking Across the World*, 270.

13. Lewis Fry Richardson, *Weather Prediction by Numerical Process* (Cambridge: Cambridge University Press, 1922; facsimile reprint, New York: Dover Publications, 1965), 219.

14. Olaf Stapledon to Agnes Miller, 8 December 1916, in Crossley, *Talking Across the World*, 192–193.

15. Olaf Stapledon, *Death into Life* (London: Methuen, 1946); reprinted in Olaf Stapledon, *Worlds of Wonder: Three Tales of Fantasy* (Los Angeles: Fantasy Publishing Co., 1949), 130 (page citation is to the reprint edition).

16. Olaf Stapledon, *The Star Maker* (London: Methuen, 1937); reprinted in *Last and First Men & Star Maker* (New York: Dover Publications, 1968), 263–264.

17. Stapledon, *Last and First Men*, 119.

18. Ibid., 117–118.

19. Ibid., 118.

20. Ibid., 129.

21. Ibid., 142.

22. Frederic W. H. Myers, *Phantasms of the Living* (London: Trübner, 1886), lxv.

23. Frederic W. H. Myers, *Science and a Future Life* (London: Macmillan, 1893), 50.

24. Stapledon, *Last and First Men*, 222.

25. Fred Hoyle, *The Black Cloud* (London: Heinemann, 1957; reprint, Harmondsworth, U.K.: Penguin Books, 1960), 158 (page citation is to the reprint edition).

26. Irving J. Good, "The Mind-Body Problem, or Could an Android Feel Pain?" in Jordan M. Scher, ed., *Theories of the Mind* (New York: The Free Press of Glencoe, 1962), 496–497.

27. Irving J. Good, personal communication, 12 July 1994.

28. Irving J. Good, "Speculations Concerning the First Ultraintelligent Machine," *Advances in Computers* 6 (1965): 35–36.

29. Paul Baran, "Is the UHF Frequency Shortage a Self Made Problem?" address to Marconi Centennial Symposium, Bologna, Italy, 23 June 1995.

30. Ibid.

31. Lewis Thomas, "Social Talk," *New England Journal of Medicine* 287, no. 19 (9 November 1973): 974.

32. Olaf Stapledon, *Nebula Maker* (Hayes, Middlesex: Bran's Head Books, 1976); reprinted in *Nebula Maker & Four Encounters* (New York: Dodd, Mead & Company, 1983), 47–48.

33. Stapledon, *Star Maker*, 332.

34. Ibid.

CHAPTER 12

1. Nathaniel Hawthorne, 1851, *The House of the Seven Gables,* centenary ed. (Columbus: Ohio State University Press, 1965), 264.

2. Loren Eiseley, *The Invisible Pyramid* (New York: Scribner's, 1970), 21.

3. William of Malmesbury, ca. 1125, in J. A. Giles, ed., *William of Malmesbury's Chronicle of the Kings of England; from the Earliest Period to the Reign of King Stephen* (London: Henry Bohn, 1847), 174.

4. Ibid., 181.

5. *The Famous History of Frier Bacon, Containing the wonderful things that he did in his Life; Also the manner of his Death, with the Lives and Deaths of the two Conjurers Bungey and Vandermast. Very pleasant and delightful to be read* (London: T. Passenger, 1679), 12–13.

6. Ibid., 15.

7. Ibid., 17.

8. Warren S. McCulloch, "Where Is Fancy Bred?" in Henry W. Brosin, ed., *Lectures on Experimental Psychiatry* (Pittsburgh: University of Pennsylvania Press, 1961), reprinted in *Embodiments of Mind* (Cambridge: MIT Press, 1965), 229.

9. Olaf Stapledon, *Nebula Maker* (Hayes, Middlesex: Bran's Head Books, 1976); reprinted in *Nebula Maker & Four Encounters* (New York: Dodd, Mead & Company, 1983), 38.

10. Robert Davidge, "Processors as Organisms," University of Sussex, School of Cognitive and Computing Science, CSRP no. 250, October 1992, 2.

11. Ibid.

12. Samuel Butler, *Luck, or Cunning, as the main means of Organic Modification?* (London: Trübner & Co., 1887); reprinted as vol. 8 of *The Shrewsbury Edition of the Works of Samuel Butler* (London: Jonathan Cape, 1924), 58.

13. W. Daniel Hillis, "New Computer Architectures and Their Relationship to Physics, or Why Computer Science Is No Good," *International Journal of Theoretical Physics* 21, nos. 3–4 (April 1982): 257.

14. Samuel Butler, *Life and Habit* (London: Trübner & Co., 1878), 128–129.

15. Olaf Stapledon, *Last and First Men* (London: Methuen, 1930); reprinted, from the U.S. edition of 1931, in *Last and First Men & Star Maker* (New York: Dover Publications, 1968), 226.

16. William H. Calvin, "Fast Tracks to Intelligence (Considerations from Neurobiology and Evolutionary Biology)," in George Marx, ed., *Bioastronomy—The Next Steps: Proceedings of the 99th Colloquium of the IAU* (New York: Kluwer Academic Publishers, 1988), 241.

17. George Dyson, *Grenade Fighting: The Training and Tactics of Grenadiers* (New York: George H. Doran Co., 1917), 11.

18. George Dyson, *Grenade Warfare: Notes on the Training and Organization of Grenadiers* (London: Sifton, Praed & Co., 1915), 6.

19. Ibid., 8.

20. Ibid., 7.

21. Ibid., 11.

22. Garet Garrett, *Ouroboros; or, the Mechanical Extension of Mankind* (New York: Dutton, 1926), 51.

23. Sir George Dyson, "Fred Devenish and Others," *R.C.M. Magazine* 51, no. 2 (1955): 36.

24. Sir George Dyson, *Fiddling While Rome Burns* (Oxford: Oxford University Press, 1954), 30–31.

25. Sir George Dyson, address to the Royal College of Music, September 1949; reprinted in Christopher Palmer, ed., *Dyson's Delight: An Anthology of Sir George Dyson's Writings and Talks on Music* (London: Thames Publishing, 1989), 80.

26. Dyson, *Fiddling*, 32–34.

27. W. Daniel Hillis, "Intelligence as an Emergent Behavior; or, The Songs of Eden," *Daedalus*, (winter 1988) (Proceedings of the American Academy of Arts and Sciences 117, no. 1), 177–178.

28. Felix Mendelssohn to Marc-André Souchay, 15 October 1842, in Paul Mendelssohn Bartholdy, ed., *Letters of Felix Mendelssohn Bartholdy from 1833–1847* (London, 1864), 23–24.

29. John Wilkins, *Mercury; or, the Secret and Swift messenger: Shewing, How a Man may with Privacy and Speed communicate his Thoughts to a Friend at any distance* (London: John Maynard, 1641), 141, 143.

30. J. B. S. Haldane, "Man's Destiny," *Possible Worlds* (New York: Harper & Brothers, 1928), 303.

31. Garrett, *Ouroboros*, 19.

32. Ibid., 24.

33. Ibid., 100.

34. Ibid., 92.

35. Ibid., 51.

36. Isaac Newton, *Opticks; or, A Treatise of the Reflections, Refractions, Inflections and Colours of Light. The Fourth Edition, Corrected* (London: William Innys, 1730); reprinted, with a foreword by Albert Einstein (London: G. Bell, 1931; New York, Dover Publications, 1952), 370 (page citation is to the 1952 edition).

37. Henry David Thoreau, "Walking," *Atlantic Monthly* 9, no. 56 (June 1862): 665.

INDEX